SPICE SCIENCE and TECHNOLOGY

FOOD SCIENCE AND TECHNOLOGY

A Series of Monographs, Textbooks, and Reference Books

EDITORIAL BOARD

Owen R. Fennema University of Wisconsin—Madison
Marcus Karel Rutgers University
Gary W. Sanderson Universal Foods Corporation
Steven R. Tannenbaum Massachusetts Institute of Technology
Pieter Walstra Wageningen Agricultural University
John R. Whitaker University of California—Davis

1. Flavor Research: Principles and Techniques, *R. Teranishi, I. Hornstein, P. Issenberg, and E. L. Wick*
2. Principles of Enzymology for the Food Sciences, *John R. Whitaker*
3. Low-Temperature Preservation of Foods and Living Matter, *Owen R. Fennema, William D. Powrie, and Elmer H. Marth*
4. Principles of Food Science
   Part I: Food Chemistry, *edited by Owen R. Fennema*
   Part II: Physical Methods of Food Preservation, *Marcus Karel, Owen R. Fennema, and Daryl B. Lund*
5. Food Emulsions, *edited by Stig E. Friberg*
6. Nutritional and Safety Aspects of Food Processing, *edited by Steven R. Tannenbaum*
7. Flavor Research: Recent Advances, *edited by R. Teranishi, Robert A. Flath, and Hiroshi Sugisawa*
8. Computer-Aided Techniques in Food Technology, *edited by Israel Saguy*
9. Handbook of Tropical Foods, *edited by Harvey T. Chan*
10. Antimicrobials in Foods, *edited by Alfred Larry Branen and P. Michael Davidson*
11. Food Constituents and Food Residues: Their Chromatographic Determination, *edited by James F. Lawrence*
12. Aspartame: Physiology and Biochemistry, *edited by Lewis D. Stegink and L. J. Filer, Jr.*
13. Handbook of Vitamins: Nutritional, Biochemical, and Clinical Aspects, *edited by Lawrence J. Machlin*
14. Starch Conversion Technology, *edited by G. M. A. van Beynum and J. A. Roels*
15. Food Chemistry: Second Edition, Revised and Expanded, *edited by Owen R. Fennema*

16. Sensory Evaluation of Food: Statistical Methods and Procedures, *Michael O'Mahony*
17. Alternative Sweetners, *edited by Lyn O'Brien Nabors and Robert C. Gelardi*
18. Citrus Fruits and Their Products: Analysis and Technology, *S. V. Ting and Russell L. Rouseff*
19. Engineering Properties of Foods, *edited by M. A. Rao and S. S. H. Rizvi*
20. Umami: A Basic Taste, *edited by Yojiro Kawamura and Morley R. Kare*
21. Food Biotechnology, *edited by Dietrich Knorr*
22. Food Texture: Instrumental and Sensory Measurement, *edited by Howard R. Moskowitz*
23. Seafoods and Fish Oils in Human Health and Disease, *John E. Kinsella*
24. Postharvest Physiology of Vegetables, *edited by J. Weichmann*
25. Handbook of Dietary Fiber: An Applied Approach, *Mark L. Dreher*
26. Food Toxicology, Parts A and B, *Jose M. Concon*
27. Modern Carbohydrate Chemistry, *Roger W. Binkley*
28. Trace Minerals in Foods, *edited by Kenneth T. Smith*
29. Protein Quality and the Effects of Processing, *edited by R. Dixon Phillips and John W. Finley*
30. Adulteration of Fruit Juice Beverages, *edited by Steven Nagy, John A. Attaway, and Martha E. Rhodes*
31. Foodborne Bacterial Pathogens, *edited by Michael P. Doyle*
32. Legumes: Chemistry, Technology, and Human Nutrition, *edited by Ruth H. Matthews*
33. Industrialization of Indigenous Fermented Foods, *edited by Keith H. Steinkraus*
34. International Food Regulation Handbook: Policy • Science • Law, *edited by Roger D. Middlekauff and Philippe Shubik*
35. Food Additives, *edited by A. Larry Branen, P. Michael Davidson, and Seppo Salminen*
36. Safety of Irradiated Foods, *J. F. Diehl*
37. Omega-3 Fatty Acids in Health and Disease, *edited by Robert S. Lees and Marcus Karel*
38. Food Emulsions: Second Edition, Revised and Expanded, *edited by Kåre Larsson and Stig E. Friberg*
39. Seafood: Effects of Technology on Nutrition, *George M. Pigott and Barbee W. Tucker*
40. Handbook of Vitamins: Second Edition, Revised and Expanded, *edited by Lawrence J. Machlin*
41. Handbook of Cereal Science and Technology, *Klaus J. Lorenz and Karel Kulp*
42. Food Processing Operations and Scale-Up, *Kenneth J. Valentas, Leon Levine, and J. Peter Clark*
43. Fish Quality Control by Computer Vision, *edited by L. F. Pau and R. Olafsson*

44. Volatile Compounds in Foods and Beverages, *edited by Henk Maarse*
45. Instrumental Methods for Quality Assurance in Foods, *edited by Daniel Y. C. Fung and Richard F. Matthews*
46. Listeria, Listeriosis, and Food Safety, *Elliot T. Ryser and Elmer H. Marth*
47. Acesulfame-K, *edited by D. G. Mayer and F. H. Kemper*
48. Alternative Sweeteners: Second Edition, Revised and Expanded, *edited by Lyn O'Brien Nabors and Robert C. Gelardi*
49. Food Extrusion Science and Technology, *edited by Jozef L. Kokini, Chi-Tang Ho, and Mukund V. Karwe*
50. Surimi Technology, *edited by Tyre C. Lanier and Chong M. Lee*
51. Handbook of Food Engineering, *edited by Dennis R. Heldman and Daryl B. Lund*
52. Food Analysis by HPLC, *edited by Leo M. L. Nollet*
53. Fatty Acids in Foods and Their Health Implications, *edited by Ching Kuang Chow*
54. *Clostridium botulinum*: Ecology and Control in Foods, *edited by Andreas H. W. Hauschild and Karen L. Dodds*
55. Cereals in Breadmaking: A Molecular Colloidal Approach, *Ann-Charlotte Eliasson and Kåre Larsson*
56. Low-Calorie Foods Handbook, *edited by Aaron M. Altschul*
57. Antimicrobials in Foods: Second Edition, Revised and Expanded, *edited by P. Michael Davidson and Alfred Larry Branen*
58. Lactic Acid Bacteria, *edited by Seppo Salminen and Atte von Wright*
59. Rice Science and Technology, *edited by Wayne E. Marshall and James I. Wadsworth*
60. Food Biosensor Analysis, *edited by Gabriele Wagner and George G. Guilbault*
61. Principles of Enzymology for the Food Sciences: Second Edition, *John R. Whitaker*
62. Carbohydrate Polyesters as Fat Substitutes, *edited by Casimir C. Akoh and Barry G. Swanson*
63. Engineering Properties of Foods: Second Edition, Revised and Expanded, *edited by M. A. Rao and S. S. H. Rizvi*
64. Handbook of Brewing, *edited by William A. Hardwick*
65. Analyzing Food for Nutrition Labeling and Hazardous Contaminants, *edited by Ike J. Jeon and William G. Ikins*
66. Ingredient Interactions: Effects on Food Quality, *edited by Anilkumar G. Gaonkar*
67. Food Polysaccharides and Their Applications, *edited by Alistair M. Stephen*
68. Safety of Irradiated Foods: Second Edition, Revised and Expanded, *J. F. Diehl*
69. Nutrition Labeling Handbook, *edited by Ralph Shapiro*
70. Handbook of Fruit Science and Technology: Production, Composition, Storage, and Processing, *edited by D. K. Salunkhe and S. S. Kadam*

71. Food Antioxidants: Technological, Toxicological, and Health Perspectives, *edited by D. L. Madhavi, S. S. Deshpande, and D. K. Salunkhe*
72. Freezing Effects on Food Quality, *edited by Lester E. Jeremiah*
73. Handbook of Indigenous Fermented Foods: Second Edition, Revised and Expanded, *edited by Keith H. Steinkraus*
74. Carbohydrates in Food, *edited by Ann-Charlotte Eliasson*
75. Baked Goods Freshness: Technology, Evaluation, and Inhibition of Staling, *edited by Ronald E. Hebeda and Henry F. Zobel*
76. Food Chemistry: Third Edition, *edited by Owen R. Fennema*
77. Handbook of Food Analysis: Volumes 1 and 2, *edited by Leo M. L. Nollet*
78. Computerized Control Systems in the Food Industry, *edited by Gauri S. Mittal*
79. Techniques for Analyzing Food Aroma, *edited by Ray Marsili*
80. Food Proteins and Their Applications, *edited by Srinivasan Damodaran and Alain Paraf*
81. Food Emulsions: Third Edition, Revised and Expanded, *edited by Stig E. Friberg and Kåre Larsson*
82. Nonthermal Preservation of Foods, *Gustavo V. Barbosa-Cánovas, Usha R. Pothakamury, Enrique Palou, and Barry G. Swanson*
83. Applied Dairy Microbiology, *edited by Elmer H. Marth and James L. Steele*
84. Milk and Dairy Product Technology, *Edgar Spreer*
85. Lactic Acid Bacteria: Microbiology and Functional Aspects, Second Edition, Revised and Expanded, *edited by Seppo Salminen and Atte von Wright*
86. Handbook of Vegetable Science and Technology: Production, Composition, Storage, and Processing, *edited by D. K. Salunkhe and S. S. Kadam*
87. Polysaccharide Association Structures in Food, *edited by Reginald H. Walter*
88. Food Lipids: Chemistry, Nutrition, and Biotechnology, *edited by Casimir C. Akoh and David B. Min*
89. Spice Science and Technology, *Kenji Hirasa and Mitsuo Takemasa*

Additional Volumes in Preparation

Complex Carbohydrates in Foods, *edited by Sungsoo Cho, Leon Prosky, and Mark Dreher*

Handbook of Food Preservation, *edited by M. Shafiur Rahman*

Coloring of Food, Drugs, and Cosmetics, *Gisbert Otterstatter*

SPICE SCIENCE and TECHNOLOGY

KENJI HIRASA
MITSUO TAKEMASA

Lion Corporation
Tokyo, Japan

CRC Press
Taylor & Francis Group
Boca Raton London New York

CRC Press is an imprint of the
Taylor & Francis Group, an **informa** business

ISBN: 0-8247-0144-5

This book is printed on acid-free paper.

Headquarters
Marcel Dekker
270 Madison Avenue, New York, NY 10016
tel: 212-696-9000; fax: 212-685-4540

World Wide Web:
http://www.dekker.com

The publisher offers discounts on this book when ordered in bulk quantities. For more information, write to Special Sales/Professional Marketing at the headquarters address above.

Copyright © 1998 by Marcel Dekker   All Rights Reserved.

Neither this book nor any part may be reproduced or transmitted in any form or by any means, electronic or mechanical, including photocopying, microfilming, and recording, or by any information storage and retrieval system, without permission in writing from the publisher.

Current printing (last digit):
10 9 8 7 6 5 4 3 2

PRINTED IN THE UNITED STATES OF AMERICA

Preface

The world spice market is expanding because of many factors including the development of world trade. The recent popularity of herbs and their increased use in the food industry seems to contribute to this trend. Spice has been utilized in numerous fields and has been a part of human life throughout history. There have been an increasing number of scientific studies on the specific functional compounds and of future effective applications of spices. This book does not merely aim to be an introduction to spice science. Basic knowledge is explained systematically and applications are covered in as much detail as possible.

Chapter 1 presents an overview of spice science, definitions of spices and herbs, and their four basic effects: flavoring, pungency, coloring, and deodorizing/masking. In Chapter 2, we outline spice specifications/standards used in both spice-exporting and -consuming nations, their basic chemical properties, and their "biological" characteristics (such as insect infections and microbial contamination). We present various cooking techniques in Chapter 3 that enhance the four basic effects mentioned in Chapter 1, and in Chapter 4 we also present "pattering theory"—a very unique way of thinking in the field of spice technology, in which the relationship of each spice to each raw material cooked and to each cooking method is summarized in charts illustrating the most efficient uses of spices. In Chapters 5 through 7, the various physiological properties of spices are described. Chapter 5 presents the physiological function of pungency, especially the effect of red pepper on the human body. Reports on spices belonging to the *Labiatae* family and their effects (thrombus formation, symptoms of stress, and more) are also introduced in this chapter. Chapter 6 discusses the antimicrobial and antioxidant properties of various spices. These properties have been studied for many years and are considered very important. Chapter 7 discusses the physiological effects of spice aromas. Aromas can influence human brain waves, sleepiness, autonomic-nerve function, and the immune system. Their effects as a sex hormone are also briefly explained in this chapter.

Because of the variety of new functions of spices mentioned in this book, spice use will surely increase not only in the cooking industry, but also in other areas. This book can cover only a fraction of spice's attributes and possibilities. We hope, however, that people who read this book will understand the many possible functions of spice and discover other unique applications themselves.

Kenji Hirasa
Mitsuo Takemasa

Contents

Preface	*iii*
1. Spices and Herbs: Basic Concepts	1
2. Spice Qualities and Specifications	29
3. Cooking with Spices	53
4. The Patterning Theory of Spice Use	85
5. Physiological Effects of Spice Components	141
6. Antimicrobial and Antioxidant Properties of Spices	163
7. The Physiological Effects of Flavor/Aroma	201
Index	*213*

1
Spices and Herbs: Basic Concepts

I. DEFINITIONS AND CLASSIFICATIONS OF SPICES AND HERBS

A. A Definition of Spice

The definition of a spice differs according to the country or region in the world. Spices are sometimes defined according to where they are grown, whether they are dry or wet, or their historical background (i.e., when they started to be used as spices). However, these definitions are not always accurate from the viewpoint of their functions and constituents. The term "spice" can be defined as the dry parts of a plant, such as roots, leaves, and seeds, which impart to food a certain flavor and pungent stimuli. Table 1.1 shows the classification of spices from a botanical viewpoint. This table shows that many kinds of Labiatae, Umbelliferae, and Zingiberaceae plants are considered as spices.

TABLE 1.1 Botanical Classification of Spices

Angiospermae			
Dicotyledoneae			
Sympetalae			
Tubiflorae		Labiatae	Mint, Marjoram, Basil, Thyme, Sage, Rosemary, Oregano, Savory
		Solanaceae	Red Pepper, Paprika
		Pedaliaceae	Sesame
Campanulatae		Compositae	Tarragon
Archichlamydeae			
Piperales		Piperaceae	Pepper
Ranales		Myristicaceae	Nutmeg, Mace
		Lauraceae	Bay Leaves, Cinnamon
		Magroliaceae	Star Anise
Rhoeadales		Cruciferae	Mustard, Horseradish
Rosales		Leguminosae	Fenugreek
Geraniales		Rutaceae	Japanese Pepper
Myrtiflorae		Myrtaceae	Allspice, Clove
Umbelliflorae		Umbelliferae	Parsley, Celery, Dill, Cumin, Anise, Fennel, Caraway, Coriander
Monocotyledoneae			
Liliiflorae		Liliaceae	Garlic, Onion, Leek
		Iridaceae	Saffron
Scitamineae		Zingiberaceae	Cardamom, Ginger, Turmeric
Orchidales		Orchidaceae	Vanilla

B. Differences Between Spices and Herbs

Many types of herbs are now used around the world. The word "herb" comes from the Labit *herba*, meaning a medical plant. The meaning of herb in a narrow sense is a nonlasting plant that withers after blooming without its stems becoming woody. Perennial herbs are widely used for the purpose of dying and gardening and other as well as for medicine and food. In Europe, herbs have long been utilized for medical purposes. Some edible herbs belong to the category of spice. Even some herbs containing poisonous components can be categorized as spices if the poisonous element can somehow be neutralized with heating and other cooking procedures. An herb is botanically classified as a perennial plant, but the meaning of spice comes from its use in cooking, not any plant classification. A spice should, therefore, be edible. In fact, no spice definitions distinguish clearly between a spice and an herb. The term spice can be broadly defined as a compound that has a pungent flavor or coloring activity, or one that increases appetite or enhances digestion. A spice is obtained from seeds, berries, buds, leaves, bark, and roots of plants growing mainly in the tropical, the subtropical, and the temperate zones. In this book, we include all edible leaves that are usually called herbs in the category of *spices* (we will call them "spice leaves" or "edible herbs") and so-called medicinal herbs which are not used for cooking in the category of *herbs*.

Various parts of plants are utilized as spices. Besides leaves, bark (e.g., cinnamon), buds (e.g., clove), fruits (e.g., allspice, nutmeg, mace, and Sansho—Japanese pepper), and other parts can be utilized. The concept of a spice can be summarized as follows:

1. Many plants utilized for spice are grown in the tropical, the subtropical, or the temperate zone.
2. Not the whole but part of the plant is effective as a spice.
3. The effect of a spice is characterized more or less by its stimulating flavor.

C. Definition of Spice Seasonings and Condiments

Spices are used to season many kinds of foods either alone or as part of a complex seasoning containing sugar, salt, or other ingredients. Spice seasonings and condiments are considered to be complex seasonings with which spices have been mixed.

1. Definition of spice seasoning

There are various definitions for spice seasoning: (1) Spice seasoning can be added to all kinds of foods, enhancing one's preference for these foods. Generally it has pungency or flavor, and is composed of, for example, salt or/and sugar or/and acid other than spice. (2) Spice seasoning is a salt, herb and spice which enhance or improve the flavor of food. (3) Spice seasoning is a mixture containing one or more spices or spice extracts that enhances the flavor of the original food and is added during processing by the food industry or during the cooking process in the home.

2. Definition of a condiment

A condiment, which is more commonly used than spice seasoning, can be defined as a mixture to be added to foods when eaten (not during cooking). A condiment is composed of one or more spices and spice extracts, enhancing the flavor of foods. Pickles, which are most often eaten as a side dish, are usually included in this category.

II. SPICE FUNCTIONS

A. Spice Components and Their Functions

Many people say that they do not like or use spices because of a medicine-like smell caused by spices added to food. In most cases, however, such experiences are due to a misuse of the spice. Saffron, for example, the most expensive spice, is essential when cooking pilf and bouillabaisse. A component called crocin in this spice gives it its yellowish-red color. Since crosin is insoluble in oil, saffron cannot be used to color oil- or fat-based foods. Paprika also imparts a yellowish-red color to food. The coloring component in this spice is β-carotene, which is soluble in oil but not in water. It is for this reason that paprika is used for coloring oil-containing foods such as margarine, salad dressing, and beef stew. As mentioned above, both paprika and saffron give a similar yellowish-red color, but the nature of each spice is quite different, and they are not interchangeable.

The appropriate spice to use in various foods has been determined by trial and error for a long time. But it is still useful to consider how each spice could be utilized in new combinations of raw materials and spices from the viewpoint of food science. Each spice contains various effective components, and it is necessary to understand the characteristics of each component and how these change during cooking in order to make full use of its effect.

B. Basic Uses for Spices

Spices have various effects when used in foods: not only do they impart flavor, pungency, and color characteristics, they also have antioxidant, antimicrobial, pharmaceutical, and nutritional properties. In addition to these direct effects of spices, complex or secondary effects can be achieved by using spices for cooking. Such effects include salt reduction, sugar reduction, and improved texture for certain foods. We use as an example here the role of cinnamon in texture improvement as an example of a spice's secondary function.

As we will explain in a later chapter, cinnamon possesses strong inhibitory activity against most microbes. Such antimicrobial activity is being investigated as a way to improve food texture. Cinnamon is sometimes used in producing bread. Yeast provides the fermentation necessary for bread to rise, but it is also necessary to avoid too much fermentation in order to increase the yield of the finished product. Cinnamon, therefore, not only lends its typical flavor to bread, but also suppresses yeast activity via its antimicrobial activity. In this case, the cinnamon flavor can be called a direct effect and the texture improvement a secondary effect. The direct and complex effects that can be obtained from spices are summarized in Table 1.2.

Spices and Herbs: Basic Concepts

TABLE 1.2 Direct and Complex Effects of Spices

Direct effect	Complex effect
Flavor	Increased appetite
Taste (pungency, bitterness, sweetness)	Masking effect
Color (red, green, yellow)	Improvement of texture
Antifungal effect	Preservation
Antibacterial effect	
Antioxidant effect	

The objectives when using spices in cooking or in processed foods should be clear. The basic effects of spices when used for cooking fall into four categories: flavoring, pungency, coloring, and deodorizing/masking. The deodorizing/masking function may overlap with the other three, but it is often desired as an effect on its own. Each spice performs at least one of these functions, as shown in Table 1.3. The use of spices is often limited by individual likes and dislikes for certain flavors. However, most spices can be used in most types of cooking. The use of all spices would be expanded greatly by understanding the cooking methods best suited to each spice and food.

We will give here a brief summary of the first three categories above. (Details about chemical compounds and their use in cooking will be explained in later chapters.) The final (deodorizing/masking) function will be fully discussed in Chapter 3.

TABLE 1.3 Basic Spice Uses

Basic uses	Spice
Flavoring	Allspice, cinnamon, basil, dill, nutmeg, fennel, parsley, anise, marjoram, cumin, mint, cardamom, mace, tarragon
Masking/Deodorizing	Garlic, clove, rosemary, onion, bay leaves, thyme, sage, coriander, caraway, oregano
Pungency	Pepper, red pepper, mustard, horseradish, Japanese pepper, ginger
Colorant	Turmeric, paprika, saffron

III. FLAVOR

A. Flavor Components

A spice is a plant composed of fiber, sugar, fat, protein, ash, gum, essential oils, and other components. Among these components, it is the volatile essential oil that imparts to a spice its particular flavor. Essential oils consist not of a single chemical compound, but of a number of compounds. Spice aromas and flavors differ subtly according to the amount of essential oil and the proportion of each compound contained in it. For this reason there may be delicate differences among different samples of the same spice. In general, the flavor characteristics of two spices belonging to the same botanical category differ according to where it is grown, harvest timing, and other factors, even if their appearance is similar. Table 1.4 lists the major components of the essential oils of some spices.

TABLE 1.4 Chemical Compounds Contained in Essential Oils of Spices

Spice	Chemical compounds in essential oil
Allspice	Eugenol, thymol, phellandrene, caryophylene, cineol, methyl eugenol
Anise	Anethole, methyl chavicol, anise aldehyde, limonene
Basil	Methyl chavicol, linalool, cineol, anethole
Bay leaf	Cineol, α-pinene, phellandrene, eugenol, linalool, borneol
Caraway	d-Carvone, limonene, carveol
Cardamom	Cineol, α-terpinyl acetate, limonene, sabinene, myrcene
Celery	Limonene, selinene, sesqueterpene alcohol
Cinnamon	Cinnamaldehyde, eugenol, caryophylene, pinene
Clove	Eugenol, caryophylene, acetyl-eugenol
Coriander	Linalool, α,β-pinene, p-cymene
Cumin	Cuminaldehyde, phellandrene, limonene
Fennel	Anethole, limonene, fenchone, α-pinene, camphene
Garlic	Diallyl disulfide, diallyl trisulfide, allyl propyl disulfide
Ginger	Gingiberene, phellandrene, borneol, linalool, shogaol, gingeroene
Marjoram	α-terpineol, terpinene, terpinene-4-ol, α-pinene, linalool, borneol, chavicol
Nutmeg/Mace	Myristisin, α-pinene, eugenol, geraniol, limonene, terpineol
Oregano	Thymol, carvacrol, α-pinene
Pepper	Piperine, caryophylene, α-pinene, phellandrene, camphene, myrcene
Rosemary	Cineol, borneol, camphor, terpineol, linalool
Sage	Cineol, linalool, camphor, borneol, α-pinene, thujone
Star anise	Anetole, methyl chavicol, α-pinene, limonene, phellandrene
Tarragon	Methyl chavicol, ocimene, myrcene, phellandrene
Thyme	Thymol, carbaclol, linalool, α-pinene, borneol
Dill	Carvone, α-pinene, limonene, phellandrene

Spices and Herbs: Basic Concepts

The major chemical components of the essential oils of spices are terpene compounds consisting of carbon, hydrogen, and oxygen. Compounds having 10, 15, and 20 carbons are called monoterpenes, sesquiterpenes, and diterpenes, respectively. A monoterpene generally has a strong aroma and is very volatile. Monoterpenes are utilized as ingredients in perfumes or essences. Green plants emit oxygen during carbon assimilation. They also release terpene compounds, which are said to strengthen the anti-cold property of spices and work as an insect repellant. Essential oils containing such terpene compounds are contained in special tissues and cells such as oil cells, oil glands, and glandular hairs. A strong aroma results and the essential oil is volatilized when these tissues and cells collapse. Figure 1.1 shows a photomicrograph of a mint leaf. An oily membrane (oval shape) contains the essential oil, and this membrane is broken to release its aromatic compounds when mint leaves are crushed or sliced.

Each spice has a typical fresh aroma, but spices used as fresh leaves also often exhibit a grassy smell. Many people dislike this grassy quality of fresh edible herbs. The flavor and aroma of some spices such as sage and thyme are actually intensified by the drying process, which eliminates not only the compounds causing the grassy smell, but also much of the moisture, resulting in a stronger flavor due to a concentration of low-volatility compounds.

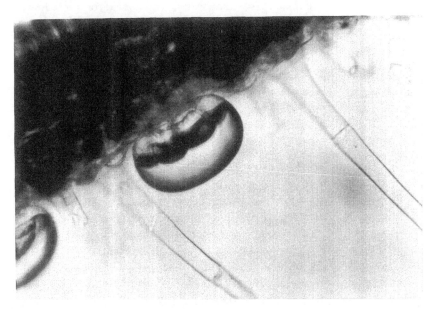

FIGURE 1.1 Microscopic picture of peppermint.

B. Flavoring Activity

Various kinds of flavor compounds are contained in spices, from those having very strong flavors to others having mild flavors. The balance of these compounds determines the total flavor of spice. Some spice compounds are resistant to heat, and others are not. With this understanding, you can change or enhance a spice's flavor with the appropriate cooking technique. As flavor compounds are somewhat volatile even at room temperature, they are certainly volatilized swiftly upon heating. Therefore, it is necessary to limit the heat used (or time heated) as much as possible to retain a spice's fresh flavor. (Likewise, increasing the heat condition will reduce a spice's flavor, if desired.)

Fresh basil will lose its aroma when cooked at high temperature, e.g., when fried in oil. On the other hand, sesame and poppy seeds, which do not have a strong aroma, become much more flavorful when roasted. Sesame seeds contain little volatile essential oil, but they do contain much nonvolatile oil, ranging from 45 to 55%. The nonvolatile fraction is composed mainly of oleic acid, stearic acid, and palmitic acid, and this oil has a slight almondlike aroma. Commercial sesame oil is manufactured by pressing roasted sesame seeds. The major flavor compounds in roasted seeds are aldehyde, acetyl pyrazine hexanal, and furfural. Red pepper contains almost no volatile essential oil, but it does contain many nonvolatile compounds. For this reason, red pepper is roasted to achieve its typical flavor. Red pepper, usually roasted, is a major ingredient in *Shichimi*, a Japanese traditional blended spice mixture. Mustard seed is also roasted for use in Indian chutney or curry.

C. Change in Flavor Compounds During Cooking

Many spices have a bitter or astringent or harsh taste, especially in their fresh state. Alkaloids, their glycosides, and organic or inorganic salts are known to have a bitter taste. An inorganic salt content exceeding 1.5% in case of spice leaves will result in a bitter taste. Furthermore, this unpleasant bitterness is stronger if potassium is a major component of the inorganic salt.

There are many undesirable taste components, including bitterness, astringency, and the above-mentioned grassy flavor. Frying in oil is one cooking technique used to decompose harshness compounds contained in spices, especially fresh leaves. Also, boiling sometimes makes spices more tasty by removing any harshness and grassy flavor when they dissolve in the boiling water.

There are some flavor compounds of spices that function to suppress unpleasant smells. The deodorizing/masking action of spice can be used to mask the unpleasant smell of fish or meat. There are also many type of unpleasant odors such as that produced upon cooking cabbage (from dimethyl sulfide), that of fish and meat (from trimethylamine and amine), and that of rancid oil (from oxidation) and the like. It is not easy to generalize about this subject because of the varied

Spices and Herbs: Basic Concepts

likes and dislikes in different cultures in addition to the numerous compounds involved. But we can say that many spices belonging to Liliaceae family have the function of masking the smell of fish and meat when cooked with them. The deodorizing function of spices will be discussed further in Chapter 3.

IV. PUNGENCY

A. Pungency Components

When tasting the essential oil of any kind of spice, a pungent sensation is perceived to a greater or lesser degree. This pungency is directly related to the concentration of the essential oil. Most such pungent sensations are considered to be due to the essential oil stimulating a taste bud on the tongue. Generally, pungent sensation perceived would be stronger for the more concentrated essential oils. There are, of course, spice compounds perceived to be pungent even at very low concentrations. As Table 1.5 indicates, the pungent principle of each spice is basically composed of different kinds of compounds. The pungent compounds contained in spices can be classified into acid-amide compounds, carbonyl compounds, thioether compounds, and isothiocyanate compounds. Each of these four groups is different in its cooking characteristics as well as its pungent sensation.

TABLE 1.5 Pungent Constituents in Various Spices

Spice	Pungent compound	Basic structure	Sensation
Red pepper	Capsaicin**, Dihydrocapsaicin**	Acid amide group R-CO-N−R \\ R	Hot
Black/White pepper	Piperine**, Chavicine**		
Japanese pepper	α-Sanshool**, β-Sanshool**		
Ginger	Zingerol*, Shogaol*	Carbonyl group	
Tade	Polygodial* (Tadenal*)	R-CO-R	
Onion	Diallyl sulfide*	Thioether group	
Garlic	Diallyl disulfide*	R-S-R	
Mustard	Allyl isothiocyanate*, P-hydroxybenzyl isothiocyanate**	Isothiocyanate group R-N-C-S	
Horseradish	Allyl-isothiocyanate*		
Radish	Butylcrotonyl isothiocyanate sulfide*		Sharp

*: Volatile; **: nonvolatile.

The chemical structures and other chemical characteristics for each pungent spice will be discussed in Chapter 2.

Pungent sensations can be divided into a "hot" sensation, which spreads into the mouth, and a "sharp" sensation, which stimulates the mucous membrane of both the nose and the oral cavity. Spices containing acid amide compounds or carbonyl compounds give the former hot sensation, and ones containing thioether compounds or isothiocyanate compounds give the latter sharp sensation. Most of the compounds that account for the "sharp" sensation are volatile compounds, while those having the "hot" sensation are generally nonvolatile. This means that most of "sharp" pungent compounds of a spice are identical with its flavor compounds, but its "hot" pungent compounds are usually different from its flavor compounds. Aromatic spices usually contain their essential oils in their cells and tissues, but some kinds of spices contain their flavor and pungency in the form of a glycoside. In these, an enzyme, which is also contained in its tissues and cells, decomposes glycoside to produce flavor and pungent compounds. The typical flavors or pungency of plants of the *Allium* genus such as garlic, onion, and leek are attributed to alkyl disulfide compounds, which are produced from alkyl cysteinsulfide by enzymatic action [1]. In the case of garlic, for example, this precursor glycoside, called alliine, exists in its plant tissue. Alliin itself does not have any smell or pungency, but easily produces allicine. Cutting one of these vegetables collapses the cells of this plant so that an enzyme called alliinase helps to decompose alliin to produce allicine.

Allicine and its analogue are not stable in air. They gradually decompose to produce disulfide or trisulfide compounds, which are the major flavor compounds found in *Allium* plants. Each spice belonging to an *Allium* plant can be characterized by the kind of major sulfide compound contained in it. There are three main R-groups to be considered for the flavor compounds of Allium plant: (1) the methyl group (CH_3-), (2) the propyl group ($CH_3CH_2CH_2$-), and (3) the allyl group ($CH_2=CH \cdot CH_2$-). Compounds with a propenyl group ($CH_3CH=CH$-) are also found in onion. Differences in smell or flavor among garlic, onion, leek, and others are due to the ratio of these three groups (Table 1.6). For example, pickled scallion

TABLE 1.6 Difference in Flavor Compounds of *Allium* Plants

	R		
	Methyl group	Propyl group	Allyl group
Garlic	12–21%	1–4	74–87
	(Ave. 13)	(Ave. 2)	(Ave. 85)
Onion	1–7%	75–96	3–9
	(Ave. 6)	(Ave. 88)	(Ave. 6)
Leek	28–91	1–67	2–9

contains mainly methyl groups, while propyl groups predominate in leek and onion. Also, the chief compounds responsible for the flavor of garlic contain allyl groups. It is generally thought that plants containing mainly methyl groups have a strong smell, while ones having propyl groups have a relatively moderate smell. Cutting an onion makes one's eyes water. The volatile compounds causing this phenomenon is produced by decomposition of (+)-S-(1-propenyl)-L-cystein-sulfoxide. Figure 1.2 shows all of the flavor precursors of *Allium* plants and their flavor characteristics [2].

B. Pungency and Cooking

Figure 1.3 shows a cross section of onion. According to studies of the distribution of flavor precursors in onion, its stalk contains twice as many as its leaves, and the interior of the leaves contains more than the exterior. This fact is very important in using onion. For example, its exterior, which is less pungent, can be used as a salad ingredient, whereas its interior is better to use for stewing (see below).

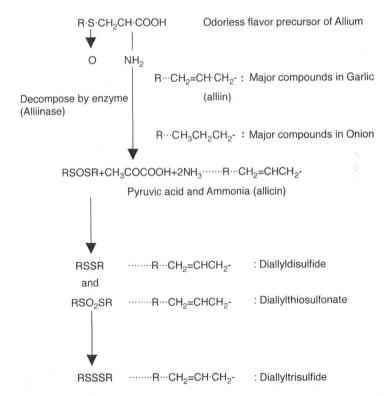

FIGURE 1.2 Flavor precursor of allium and the formation of aromatic compounds.

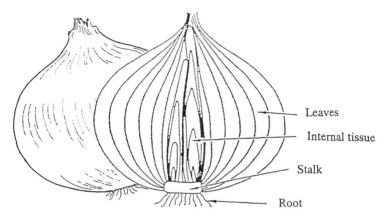

FIGURE 1.3 Cross section of onion.

Flavor or pungent characteristics can be accentuated by bringing flavor precursors in contact with an enzyme existing in onion tissue. It is for this reason that chopping an onion fine will enhance its typical pungency compared to simple slicing. The precursors of onion flavor do not have any flavor, but do have a sweet taste. Also, heating can easily deactivate the enzyme contained in onion tissues. Heating can, therefore, stop decomposition of the precursor and reduce the odors of garlic and onion.

Pungent compounds of onion when combined with protein will neutralize meat smell. It is, therefore, most effective to cook onion together with meat rather than to add it later. In addition to Lilaceae plants (including *Allium*), similar pungent characteristics can be found in plants belonging to the Compositae family and the Leguminosae family.

The pungent compounds of mustard, horseradish, and radish belong to the isothiocyanate group, and are contained in the form of glycosides in the plant tissue of their roots and leaves. This glycoside does not itself have any pungency, but when decomposed by an enzyme called myrosinase produces an isothiocyanate compound exhibiting pungency. Pungency in members of the Cruciferae family is produced by enzyme action upon glycosides in the same way as in onion and garlic. It is advisable to grate these finely so that the enzyme and glycosides are mixed well. Generally, it takes some time to decompose the glycosides of mustard or horseradish, therefore, it takes a while after grating for the pungent sensation to develop. Acid or salt as well as heat will reduce the pungency of these spices as they inhibit the glycoside decomposition. Rice vinegar is sometimes added to grated horseradish, mustard, and radish in Japanese foods for the purpose of making them less pungent.

Spices and Herbs: Basic Concepts

Onion, leek, and radish acquire a sweet taste when they are boiled. The main reason for this phenomenon is that pungent sulfide compounds are deoxidized to produce sweet mercaptan compounds.

Ginger, red pepper, and black/white pepper are considered among the "hot" pungent spices, and the pungent compounds contained in them are relatively heat-stable compared to those found in onion or mustard. They can, therefore, be used for cooking requiring heat. Ginger, whose pungency is due to volatile carbonyl compounds, is a little less stable than red and black/white pepper, whose pungency is due to nonvolatile acid amides. Red pepper, which is the most stable against heat, has little aroma because it does not contain much essential oil. Mustard and horseradish are not suitable for use in cooking requiring heat because their pungent compounds usually disappear and sometimes change into bitter compounds. The Japanese pepper *Sansho*, which has a "numbing" type of pungency, is popular in Japan and Korea where it is often used for its aroma rather than its pungency, expecting its deodorizing effect on fish. Characteristics of the pungent spices are summarized in Table 1.7.

V. COLORING AGENTS

A. Coloring Compounds in Spices

Many kinds of coloring compounds contained in spices are used in the field of food processing as natural colorants. To use spices for their coloring characteristics, it is essential to understand each coloring component and its tint. Coloring components of spice include carotenoids, flavonoids, chlorophyll, among others. Each coloring component is described in the following sections.

1. Chlorophyll

The green color of plant leaves is mainly due to chlorophyll. A bright green tint can be obtained using green leaves as they are. But it is necessary to shorten the cooking time when doing so since chlorophyll will turn brown upon heating. Often green herbs are added at the end of the cooking process to retain the green color. This also prevents flavor loss caused by heating.

2. Carotenoids

Carotenoids are oil-soluble and therefore suitable to use as a coloring agent in salad dressing. They can be used in cooking because of their heat-resistant nature. There are many types of carotenoids, e.g., β-carotene, capsantin, capsorbin, and lutein, found in paprika.

3. Flavonoids

Both anthocyanin and flavonoids are water-soluble. Anthocyanins are red, blue, purple, and pink in color, whereas flavonoids are yellow and yellowish-white.

TABLE 1.7 Flavor Characteristics of Pungent Spices

	Pungent characteristic						
	Pungent sensation	How soon pungency appears	Durability of pungency	Antiheat property	Flavoring	Deodorizing/ Masking	Coloring
Red pepper	Hot	Late	⊙	⊙	×	×	⊙
Black/white pepper	Hot	Mid-late	○–⊙	⊙	○	○	△
Japanese pepper	Hot	Late	△	○	⊙	○	×
Ginger	Hot	Mid-late	△–○	△–○	○	○	×
Tade	Hot	Mid	×–△	×	△	×	×
Onion Garlic Mustard	Sharp	Early	×	×	⊙	⊙	×
Horseradish Radish	Sharp	Early	×	×	⊙	⊙	×

Strength of effect: ⊙ > ○ > △ > ×.

Anthocyanin is also soluble in acids. When perilla, an hibiscus containing anthocyanin is immersed in vinegar, the solution becomes red. Anthocyanin pigments become stable and give a bright color when they bind with aluminum or iron. It is for this reason that a nail is often put in pickled vegetables to stabilize their bright color.

The color components found in spices are summarized in Table 1.8.

B. Using Spices as Coloring Agents

The coloring effect of spice can be manipulated by using the appropriate cooking method. The easiest way to use a spice to color a food is without dissolving it. The spice can be used as is or after crushing into a paste. For example, you can enjoy green color of parsley just by sprinkling it on a dish, but it can also be boiled and used to color a soup or sauce, although in this case its color becomes a dull green.

The tone of a spice's color changes, depending upon how it is dissolved in water or oil. You can, for example, acidify the water in order to change a spice's color tone. The color component contained in hibiscus belongs to the anthocyanin family. This color appears red when it is protonated (acidified), but it becomes colorless when the pH is raised. This component is also stabilized as it coexists with metal. The yellow color component contained in chestnut and sweet potato, which belongs to the flavonoid family, becomes bright when it is boiled with an alum because the aluminum in alum bonds to this component to stabilize the yellow color of flavonoid.

TABLE 1.8 Color Components in Spices

Color component	Tint	Spice
Carotenoid component		
β-Carotene	Reddish-orange	Red pepper, mustard, paprika, saffron
Cryptoxanthin	Red	Paprika, red pepper
Lutein	Dark red	Paprika, parsley
Zeaxanthin	Yellow	Paprika
Capsanthin	Dark red	Paprika, red pepper
Capsorbin	Purple red	Paprika, red pepper
Crocetin	Dark red	Saffron
Neoxanthin	Orange-yellow	Parsley
Violaxanthin	Orange	Parsley, sweet pepper
Crocin	Yellowish-orange	Saffron
Flavonoid	Yellow	Ginger
Curcumin	Orange-yellow	Turmeric
Chlorophylls	Green	Herb

The main points made about spice color so far can be summarized as follows:

1. Because there is a water-soluble component or an oil-soluble component in each spice, you have to choose the appropriate type to suit your cooking purpose.
2. Because some spice color components change their color tones due to pH of the solution, you may change the pH of solution in which the spice is dissolved.
3. Some spice color components will be stabilized when used with metals.
4. Because most of the color components of spice change their color tone upon heating, heating conditions must be carefully controlled.
5. For spices whose color component do not seep out of the spice during cooking, you can use it directly when cooking.
6. You can enjoy the bright green color of fresh leaves by straining them to add to sauces.
7. You can also enjoy a color tone of dried spice such as parsley and green pepper when these are added to sauce or soup.

C. Carotenoid Compounds

Recently, a nutritional value of carotenes has attracted worldwide attention. Some spices contain carotene compounds, as detailed below.

Carotene was extracted from carrot (*Daucus carota*) by Wackenroder in Germany for the first time in 1831. It is for this reason that this compound was named carotene. Carotene is contained mainly in plants including vegetables. Carotenoids include hydrocarbon compounds, called carotene, and their oxygenated derivative, called xanthophylls. Some microbes and plants can synthesize carotenoids, but animals cannot. Therefore, a carotenoid obtained from an animal can be assumed to be derived from carotenoids contained in livestock feed. The beautiful red color of a goldfish and a varicolored carp is due to the carotenoid group. Likewise, flamingo, shrimp, and sea bream derive their color from carotenoids they ingest.

There are a variety kinds of carotenoid compounds available. Carotenoid is contained chiefly in carrot or provitamin A, which is converted to vitamin A in the human body. Although excess intake of vitamin A is thought to be harmful, carotenoids have no harmful effect, even when accumulated in the body. Table 1.9 shows that a large amount of provitamin A is contained in paprika. β-Carotene, which is the most popular provitamin A, is also widely used in the fields of medical drugs, cosmetics, food, and feed. In the field of food, it has been used as a colorant in cheese, butter, margarine, ice cream, and some confectionery. Its use has increased in the field of nutritional foods and drugs because of its nutrient value. The effectiveness of carotenoid will be understood more in the future, and their use and application will likely be increased in all of the above fields.

Spices and Herbs: Basic Concepts

TABLE 1.9 Carotene Contents in Paprika and Other Brightly Colored Vegetables

Vegetable	Carotene (μg)	Vegetable	Carotene (μg)
Paprika powder	20,000	Asatsuki	2300
Perilla	8700	Mustard greens	2300
Parsley	7500	Watercress	1800
Carrot	7300	Turfed stone leak	900
Garland chrysanthemum	3400	Leaf-blade green	860
Chinese chive	3300	Mitsuba	790
Spinach	3100	Leek	610
Japanese radish	2600	Sweet pepper	270

More than 1000 kinds of carotenoids have been identified thanks to the rapid development of analytical techniques. α-, β-, γ-Carotene, lycopene, zeaxanthin, lutein, capsantin, cantaxanthin, astaxanthin, and others are among them. Each of these carotenoids, when eaten by an animal, is metabolized to yield its typical color. A prawn, for example, metabolizes β-carotene, zeaxanthin, and astaxanthin to yield its color. Red pepper and paprika contain many carotenoid compounds. The main carotenoid compounds of these spices are said to be capsantin, lutein, zeaxanthin, β-carotene, and cryptoxanthin, but some reports say that 20–30 kinds of carotenoid compounds are included in these spices. The kind and amount of carotenoid compounds varies according to the kind of red pepper and where they are grown. Fish and seafood can be given color by feeding them carotenoid-containing spices. The kind of carotenoid involved differs from one kind of seafood to another. See Table 1.10 for a partial list of such colorant effects. A kind of carotenoid contained in a red sea bream is shown in Figure 1.4. In many countries of Europe, North America, and South America, astaxanthin, which can be obtained through oxidation metabolism of carotene, is added to livestock feed for Pacific salmon, silver salmon, rainbow trout, and Arctic char for the purpose of improving their color. This technique is applied to "black tiger" harvested in southeast Asia, and this color-improved "black tiger" is now being imported to Japan. Recently, some beer and wine companies in Japan are doing research on livestock feed containing yeast in which some colorant is included for the purpose of giving red color to hatchery sea bream. This type of livestock feed consists of fish meal and flour in addition to astaxanthin. It has been reported that black sea bream weighing 800–1 kg become reddish in 2–4 months on this feed. A wild sea bream becomes red when it eats shrimp and small fish that contain red colorant. However, since livestock feed contains almost no colorant, hatchery fish slowly acquire a black color. Hatchery fish tend to get a kind of sunburn near the head and lose their market value because they swim in shallow water and are exposed to the sun. Black hatchery fish can regain a bright red color when fed a carotenoid such

TABLE 1.10 Carotenoid Compounds
Determining Body Color of Fish

Fish	Carotenoid	Spice
Prawn	β-Carotene	Red pepper
	Zeaxanthin	Mustard
	Astaxanthin	Paprika
		Saffron
Varicolored carp	Zeaxanthin	Paprika
Goldfish	Rutin	Parsley
	Astaxanthin	
Young yellowtail	Zeaxanthin	Paprika
	Astaxanthin	
Salmon	Astaxanthin	
Red sea bream		
Trout		
Char		

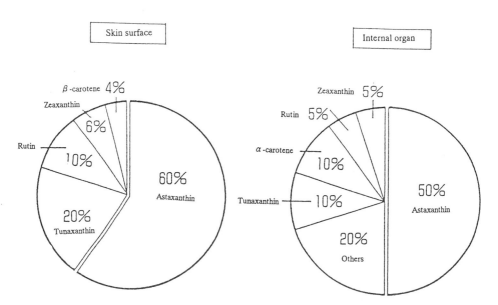

FIGURE 1.4 Carotenoid composition, both at skin surface and in internal organs, or red sea bream.

Spices and Herbs: Basic Concepts

as astaxanthin. In this case, it changes the inside meat as well as its surface. It is also reported that the hatching rate of a reddish sea bream is higher than that of a black sea bream. Carotene is also confirmed to enhance the rate of breeding caw, and is designated as an additive for livestock feed.

D. Spice Extracts as Natural Coloring Agents

Recent technological advances have enabled us to conduct research on extracting color components from spices and utilizing them as natural colorants. The following are some spices whose extracts are used in the food industry.

1. Turmeric

Botanical name: *Curcuma longa* L.
Extract method: Extract with ethanol, lipid, or organic solvent
Coloring component: Curcumin (Fig. 1.5)

Curcumin is soluble in ethanol, propylene glycol and acetic acid, and not soluble in water and ether. Its melting point is 183°C. It appears yellow at pH levels from acid to neutral, and appear reddish-brown in alkaline pH. It stains protein well, and does not reduce much. It is relatively stable against heat, but not against light, and becomes dark in the presence of metal, especially iron. Turmeric powder, turmeric oleoresin, alcohol or propylene glycol solvent are used for confectionery, agricultural products, and processed marine products.

2. Paprika and red pepper

Botanical name: *Capsicum annum* var. *cuneatum* PAUL.
Extract method: Extract with lipid or organic solvent
Coloring component: Capsanthin (Fig. 1.6)

Capsanthin is an oil-soluble component that belongs to carotenoid family and is red to orange in color. It has relatively strong red color compared to other carotenoids. It is soluble in alcohol and oil, but not in water. It does not change its color in accordance with solvent pH, and is relatively stable against heat. Although it is slightly unstable to light, its stability can be improved by adding

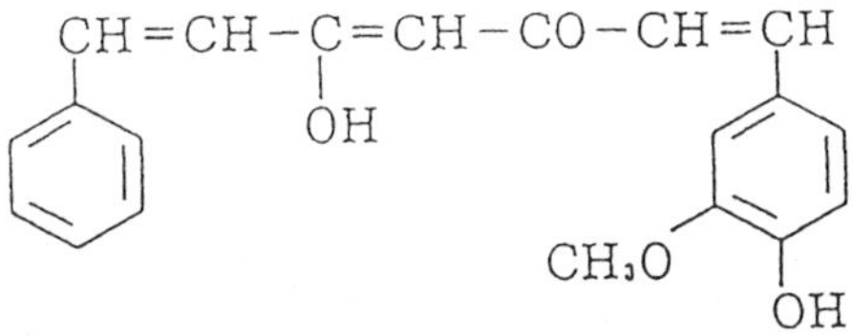

$$CH=CH-C=CH-CO-CH=CH$$
$$\quad\quad\quad\quad | $$
$$\quad\quad\quad\quad OH$$

$$CH_3O$$
$$\quad\quad OH$$

FIGURE 1.5 Chemical structure of curcumin ($C_{21}H_{20}O_6$).

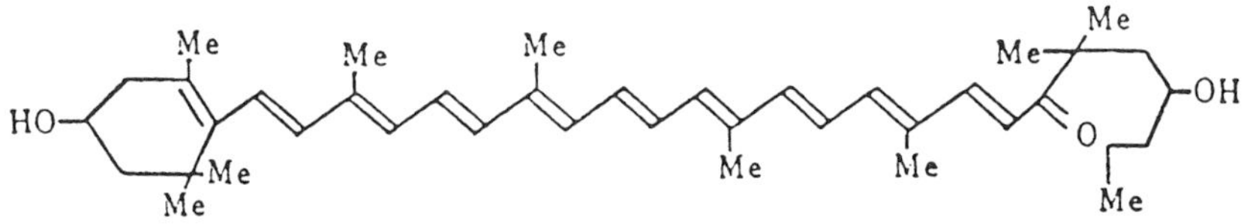

$C_{40}H_{52}O_4$

Me : CH_3

FIGURE 1.6 Chemical structure of capsanthin.

antioxidant reagent. Paprika oleoresin and emulsified paprika are also used widely for confectionery, agricultural products, and processed marine products.

3. Saffron and gardenia

Botanical name: *Crocus sativus* L. (saffron)
Fruit of *Gardenia augusta* MERR. var. *grandiflora* HORT. (gardenia)
Extract method: Extract with ethanol for saffron; extract with water or ethanol or hydrolysis for gardenia
Coloring components: Crocin and crocetin (Figs. 1.7, 1.8)

Both crosin and crocetin, which belong to the carotenoid family, are water soluble, unusual for the natural world. They have a very beautiful yellow color, but they cannot be used in beverages because they are not stable against acid or light. However, their resistance to acid can be improved by adding an antioxidant reagent. They tend to fade in the presence of iron and copper ions. Gardenia powder and its water solution are used widely for confectionery, agricultural products, and processed marine products.

4. Red perrila

Botanical name: *Perilla fruitescens* BRI TT. var. *acuta* KUDO
Extract method: Extract with ethanol
Coloring component: Sisonin, peranin

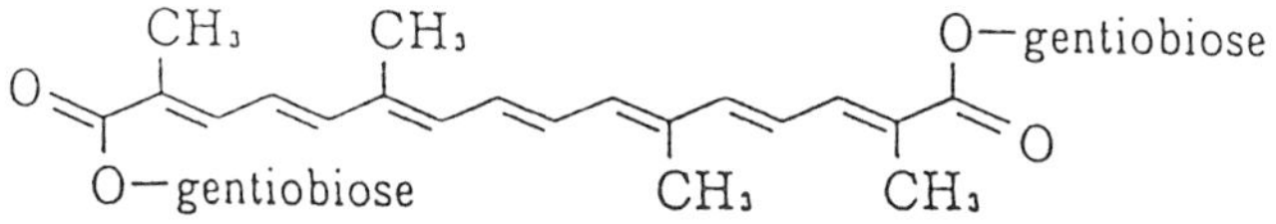

$C_{44}H_{64}O_{24}$

FIGURE 1.7 Chemical structure of crocin.

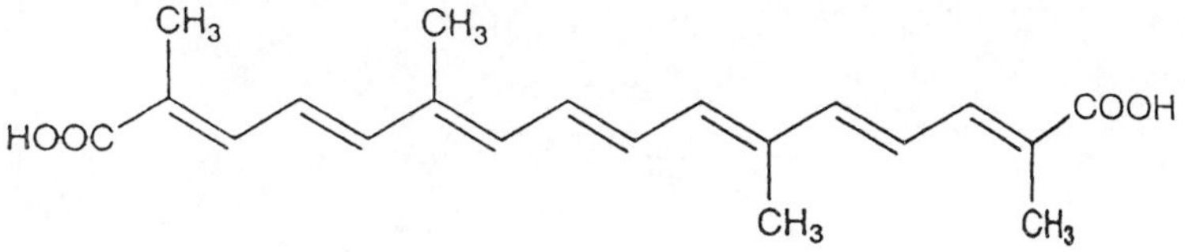

Spices and Herbs: Basic Concepts

$$C_{20}H_{24}O_4$$

FIGURE 1.8 Chemical structure of crocetin.

Both sinonin and peranin, which belong to the anthocyanin family, are soluble in water, alcohol, and acetic acid, but not in oil. They have a red color at acid pH, but change to purple and to blue when pH becomes to neutral and alkaline, respectively. However, their purple color at neutral pH and blue color at alkaline pH are not stable and easily fade and turn brown in a short period of time. They are resistant to both heat and light in acid pH. Red perilla powder and its water solution are used widely in confectionery, agricultural products, and dessert products.

5. *Onion*

   Botanical name: *Allium cepa* L.
   Extract method: Extract with ethanol or water
   Coloring component: Quercetin

   Quercetin, which belongs to the flavonoid family, is brown to reddish-brown in color. This color tone does not change greatly from neutral to alkaline pH. It is soluble in water and in alkaline solution, but tends not to be soluble in acid solution, and it is not soluble in oil. It is stable against heat and light, and has a superior dying effect. The powdered and liquid products are used for confectionery, stock farm products, and processed marine products.

6. *Hibiscus*

   Botanical name: *Hibiscus sabdariffa* L.
   Extract method: Extract with water from calyxes
   Coloring components: Anthocyanins

   Hibiscus is soluble in water and alcohol, not in oil. Its color tone is bright red at acid pH, but changes to purple at neutral and alkaline pH. It is not stable against heat and light, and is used as a pink color in soft drinks and alcoholic beverages.

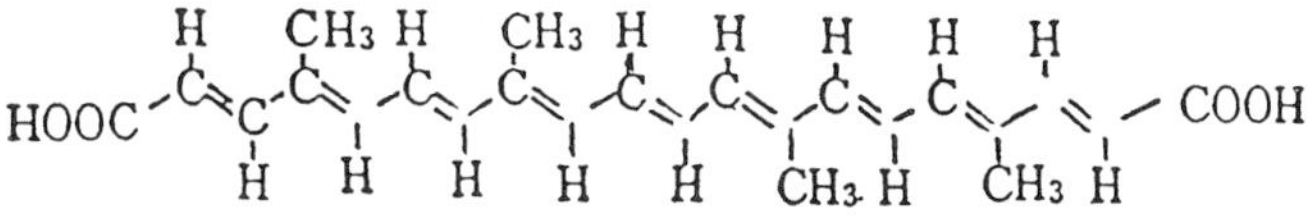

HOOC−C(H)=C(CH₃)−C(H)=C(H)−C(CH₃)=C(H)−C(H)=C(H)−C(H)=C(CH₃)−C(H)=C(H)−C(CH₃)=C(H)−C(H)=C(H)−COOH

$C_{24}H_{28}O_4$

FIGURE 1.9 Chemical structure of norbixin.

7. Annatto

Botanical name: *Bixa orellana* L.
Extract method: Extract with organic solvent or oil from annatto seed
Coloring component: Bixin, norbixin (Fig. 1.9)

Both bixin and norbixin belongs to the carotenoid family. Bixin, whose melting point is 198°C, is slightly unstable against light, and is soluble in oil and alkali solutions, but not in water. Norbixin, whose melting point is over 300°C, is soluble in acetic acid and pyridine, but not in oil. It has resistance to heat, but is slightly resistant to light. Oil solvent solutions of bixin and alkali solutions and powder products of norbixin are used widely for confectionery, stock farm products, and processed marine products.

8. Others

Western Madder The color component called alizarin (Fig. 1.10) is extracted from a root of this plant. It is an oil-soluble orange colorant, and mainly used for oily products. The color component called purpurin (Fig. 1.11) can be obtained from the root after it is held in storage for some period of time, but not from fresh root. This colorant is reddish-purple.

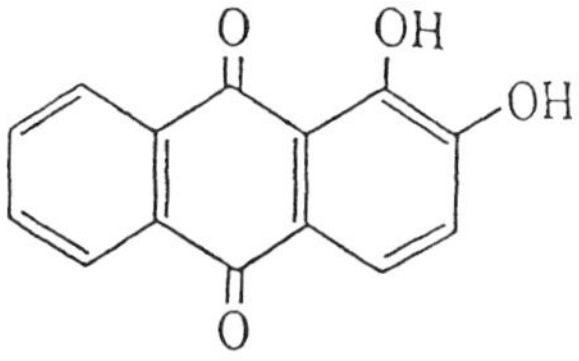

$C_{14}H_8O_4$

FIGURE 1.10 Chemical structure of alizarin.

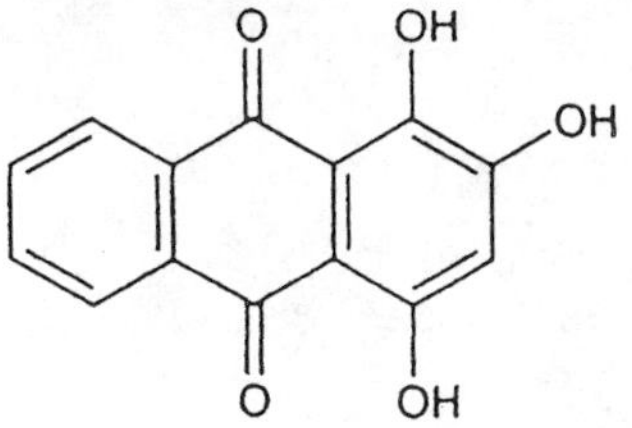

Spices and Herbs: Basic Concepts

$C_{14}H_8O_5$

FIGURE 1.11 Chemical structure of purpurin.

Japanese Indigo Plant (Smartweed Family)

The color component called indigo (Fig. 1.12) is extracted from this plant's leaves. It is used as a blue colorant and is used in fields other than food processing.

Litmus Moss The coloring component called orchil is obtained from the extract of this plant, and is mainly used as a bluish-purple colorant for beverages.

Cacao (Chinese Parasol Tree Family) Cacao bean is known as the main ingredient in chocolate. A brown colorant can be obtained from the outer skin of this bean, and is used for some candy, cold confectionery, and other products.

European Grape A purple colorant is contained in the outer skin of this grape. This colorant is extracted from the residue obtained after squeezing grape juice. It is used for jam, alcoholic beverages, and soft drinks.

Green Plant Leaves Chlorophyll can be extracted from many types of green leaves and is used widely as a green colorant.

Red Beet The color component called betanin (Fig. 1.13) can be obtained from beet root. This is a water-soluble red colorant, which is used for confectionery, jelly, or other purposes.

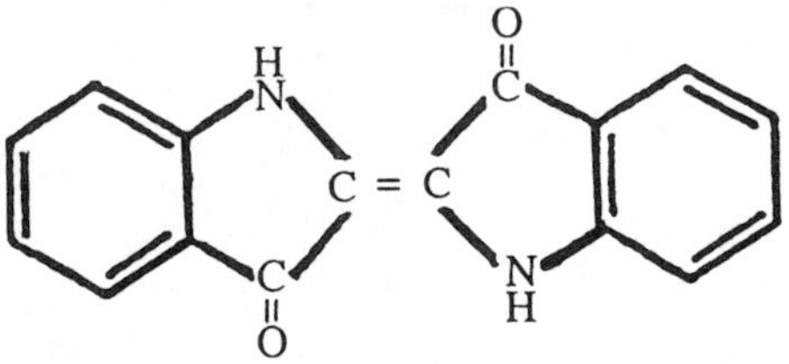

$C_{16}H_{10}N_2O_2$

FIGURE 1.12 Chemical structure of indigo.

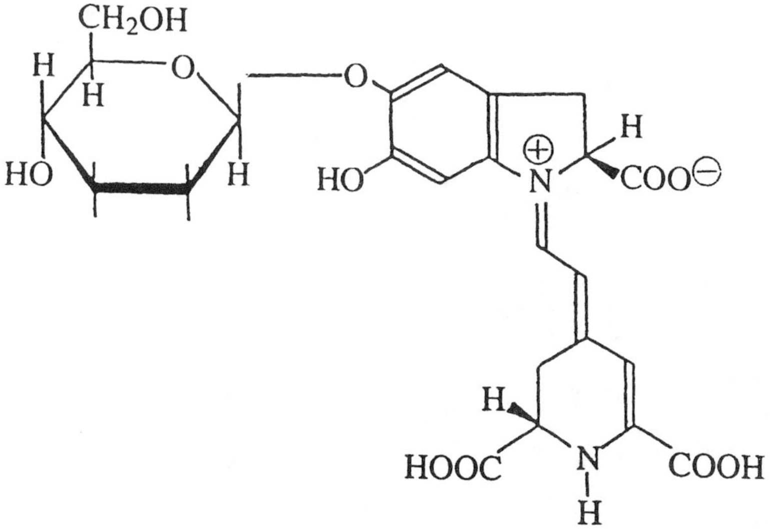

$C_{24}H_{26}N_2O_{13}$

FIGURE 1.13 Chemical structure of betanin.

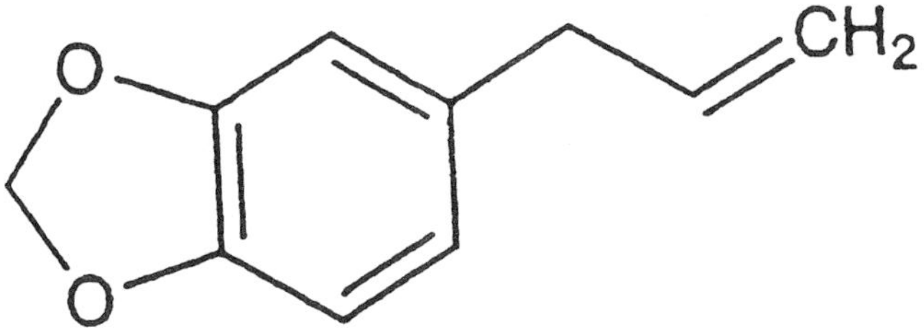

$C_{10}H_{10}O_2$

FIGURE 1.14 Chemical structure of safrole.

TABLE 1.11 Properties of Various Colorants

Color tone	Colorant	Solubility			A	B	C	D	E	F
		Water	Alcohol	Oil						
Yellow–Orange	Annatto	△	○	○	△	△	△	○	○	⊙
Yellow	Gardenia yellow	⊙	○	×	△	△	○	○	○	⊙
Orange–Red	Paprika	×	△	⊙	△	○	○	○	○	×
Yellow–Orange	β-Carotene	×	△	⊙	○	○	○	○	○	×
Orange–Reddish-purple	Cochineal	⊙	×	×	⊙	⊙	⊙	○	×	△
Orange–Red	Lack	△	⊙	×	⊙	⊙	⊙	○	×	△
Yellow–Reddish-purple	Madder	△	⊙	–	⊙	⊙	⊙	○	×	△
Purplish-red–Reddish-purple	Red cabbage	⊙	○	×	○	○	○	○	×	△
Purplish-red–Reddish-purple	Grape pericarp	⊙	○	×	○	○	○	○	×	△
Red–Reddish-purple	Grape juice	○	○	×	○	○	○	○	×	△
Red–Reddish-purple	Purple maize	○	○	×	○	○	○	○	×	△
Red–Reddish-purple	Raspberries	○	○	×	○	○	○	○	×	△
Red–Reddish-purple	Perilla	○	○	×	△	△	○	○	×	△
Yellow	Safflower yellow	⊙	○	×	○	○	○	○	△	⊙
Red	Safflower red	×	×	×	△	×	○	○	△	○
Brown	Cacao	⊙	⊙	×	⊙	⊙	⊙	○	○	⊙
Brown	Kaolang	⊙	⊙	×	⊙	⊙	⊙	○	○	○
Brown	Onion	⊙	⊙	⊙	⊙	⊙	⊙	○	○	⊙
Green	Chlorophyll	×	×	○	×	×	△	○	○	△
Yellow	Turmeric	△	△	○	△	×	○	○	○	⊙
Red	Red beet	⊙	△	×	△	○	○	○	△	○
Orange-red–Red	Monascus	○	△	×	△	×	○	○	○	△
Blue	Gardenia blue	⊙	△	×	△	○	○	○	○	⊙
Blue	Spirulina	⊙	△	×	△	×	△	○	×	△
Brown	Caramel	⊙	⊙	×	⊙	⊙	⊙	○	○	×

⊙: Very good; ○: good; △: little change; ×: great change.
A: Light-resistant; B: heat-resistant; C: salt-resistant; D: Microbial-resistant; E: Metal-resistant; F: Colorant.
Source: Ref. 3.

Safflower The yellow color components called safflower yellow (Fig. 1.14) and carthamin are obtained from the safflower flower which is reddish orange in color. This colorant is mainly used for coloring noodles and fruit juice. The red color generated by oxidation of the yellow color of this flower, called "beni", used to be utilized as a dyestuff for cloth.

The colorants mentioned above are only some of those used worldwide. Colorants and their characteristics are summarized in Table 1.11 [1]. It is expected that other colorants that can be extracted from other beautifully colored spices will be put to use with the progress in technology.

E. Stability of Natural Colorants

Spice colorants are not very stable, compared to artificial colorants. Oxygen and ultraviolet light especially tend to fade their color or oxidize them. Paprika or red pepper in which carotenoid compounds account for their color are said to suffer from oxidation especially easily. Therefore, when these colorants are used in

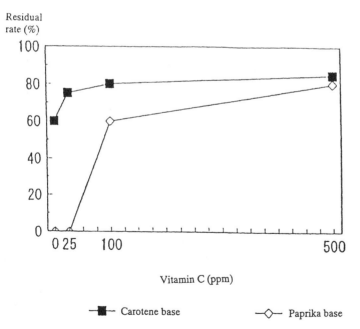

FIGURE 1.15 Synergistic effect of rutin and vitamin C on stability of carotenoid colorants. This figure shoes the residual rate of each carotenoid colorant containing rutin (0.05%) and vitamin A after 8 hours under a feed meter. (From Ref. 3.)

processed foods, it is essential to pack these products in light-impermeable materials or in packages in which inactive nitrogen is substituted for air. Also, heat can sometimes damage their color. Important factors in choosing a colorant for food are: how attractive the colorant will make the food, how natural the food color will be, and how stable the food color will be.

Many different situations can exist when a food is stored, sold, and used in the home. Food may be displayed in a light-filled shop window or may be stored at high temperature for a period of time. It is for these reasons that antioxidants are sometimes added to colorant formulations.

Antioxidants can sometimes be used to stabilize color. Figure 1.15 illustrates the stability of carotenoid compound against light [3]. Carotenoid compounds become more stable when vitamin C and enzyme treated rutin added to this compounds because vitamin or enzyme treated rutin works as anti-oxidant for carotenoids. Lease and Lease observed the effect of other antioxidants such as BHA, BHT, and α-tocopherol on the retention of the red color of ground paprika [4] and found that BHA was effective at a concentration of 0.1%, whereas BHT and α-tocopherol were not effective even at a concentration of 0.5%.

REFERENCES

1. K. Iwai and E. Nakatani, *Kosinryo Seibun no Syokuhinkino*, Koseikan, Tokyo, 1989, p. 165.
2. T. Moroe, *Shokuhin to Koryo*, Tokai Dagaku Shuppankai, Tokyo, 1979, p. 144.
3. K. Nishiyama, *Food Chem.*, 11(11): 47 (1995).
4. J. G. Lease and E. J. Lease, *Food Technol.*, 10: 403 (1956).

2
Spice Qualities and Specifications

I. SPICE SPECIFICATIONS
A. Introduction

A spice can be defined as the dried aromatic parts of natural plants, whose characteristics such as color and constitution may vary depending on year of harvest and place of harvest, among other factors. The quality of processed spices can also vary due to differences in separation and milling processes used. For these reasons it has been deemed necessary to establish quality standards or specifications for spices. Although there are no unified standards or specifications worldwide, nations that export spices often have their own quality standards to maintain their own reputations, while nations importing and consuming spices establish specifications for the purpose of consumer safety.

The quality standards most used as international guidelines are those of the

American Spice Trade Association (ASTA) and the U.S. Federal Specifications: Spices, ground and whole, and spice blends. The International Organization for Standardization established its own standards in 1969 for the quality of various spices, and there are also specifications for spices imported into and consumed in the United Kingdom and Canada. Many nations such as India and Malaysia, which are major spice-exporting nations, have their own exporting specifications in which, for example, the quality grade is classified according to the amount of extraneous matter, moisture content, etc. This chapter will discuss some major specifications for spices in both spice-consuming and spice-exporting nations.

B. Specifications in Spice-Consuming Nations

1. Specifications of the American Spice Trade Association

These are specifications for unprocessed spices imported into the United States, including edible herbs. The specifications were published first in 1969, followed by several revisions through 1975 [1]. The ASTA specifications are now being used in many other nations, including Japan, as a kind of world standard. They refer specifically to the cleanliness of the spice product, placing limits on extraneous matter (e.g., insects, insect excrement, stones, stems, sticks, etc.). Imported spices not meeting these specifications should be reconditioned at the port of entry, whereas domestic spices can be reconditioned before they are processed to be used in a consumable product.

Amounts of rodent and other animal excrement are specified by weight in the ASTA guidelines. Spices in which a certain number of insects are found alive are required to be reconditioned (e.g., fumigated). Mites and psocids, whose numbers are confirmed by a flotation test that causes these insects to float in a heated organic solvent, must not exceed the standard. Spices are not acceptable "if mold is present as expressed by percent by weight of the total number of sub-samples in excess of the specified values" or "if the total sample quantity exceeds the specified values expressed as percent by weight of insect bored or otherwise defiled seeds, leaves or roots." Also, light berry content of black pepper, though not considered extraneous matter, should not exceed 4% by weight.

The standard for extraneous matter is specified for 33 kinds of spice on the basis of actual past results. In addition to these general standards, sampling procedures and testing methods for extraneous matter are also specified. The number of samples drawn must be equal to the square root of the containers in the lot; the sample size should be one pound for high-density items and a well-filled two-pound paper bag for low-density items. Testing methods such as shifting and hand-picking for extraneous matter and light berry determination are described for black and white pepper. For nutmeg, there are the detailed guidelines for inspecting for molds and insects, which may invade through broken surfaces and cause mold contamination.

2. U.S. Federal Specifications for ground and whole spices

Besides ASTA, which regulates the cleanliness of unprocessed spices, there are U.S. federal specifications which establish quality standards for both ground and whole spices. In this U.S. specification, total ash, acid-insoluble ash, volatile oil, moisture, and other characteristics are standardized for major spices such as allspice, cardamom, coriander, cumin, cinnamon, black pepper, nutmeg, mustard, and ginger. They also establish a standard for "defective lot" according to the amount of stems or other contaminants found in certain spices. For example, bay leaves or basil leaves containing 3% or more stem material are considered to be defective according to these specifications.

3. Spice quality specifications

Specifications established by the International Organization for Standardization (ISO) in 1969 establish quality standards for 68 items, using extraneous matter, moisture content, total ash, and other chemical characteristics [3,4]. In the United Kingdom, the British Standard was established in early 1970s based on ISO specifications, specifying the amount of extraneous matter and mandating limits mainly for total ash, acid-insoluble ash, volatile oil, and moisture for several major whole spices. It also specifies the limit of crude fiber for some ground spices. Canada has a quality standard specifying total ash and acid-insoluble ash for more than 20 kinds of spices (Canadian Specification for Spices, Dressings and Seasonings). The amounts of salt and anticaking agents that can be used for spice seasonings such as celery salt and garlic salt are also regulated in this specification. For example, the amount of salt to be used in celery salt, onion salt, and garlic salt is limited to less than 75%. Anticaking agents in onion salt are limited to 2%. Japan, one of the biggest spice-consuming nations, has no quality specifications for spices, but import of spices treated with ethylene oxide or contaminated with a certain amount of aflatoxin is prohibited under the Japanese Sanitation Law.

C. Specification of Spice-Exporting Nations

Most spice-exporting nations such as India have their own exporting specifications, which also regulate the related testing methods.

1. The Indian Standards Institution

The Indian Standards Institution states quality standards for 36 kinds of both unprocessed and processed spices, ranging from major exported items such as celery, coriander, cumin, fennel, fenugreek, and turmeric to particularly Indian such as Ajowan seed and Kokun. These specifications mainly regulate the maximum moisture content. They include sampling methods and testing methods.

2. Directorate of Marketing and Inspection, administering quality control and preshipment inspection

The Government of India has prescribed standards for almost all exported spice items and graded each item using "Agmark" grades [5]. The kinds of spices include unprocessed spices such as cardamom, celery, coriander, cumin, fennel, fenugreek, ginger, black pepper, and turmeric as well as ground spices such as coriander, cumin, curry powder, fennel, fenugreek, ginger, black pepper, and turmeric.

Grade specifications are established for age-old, familiar trade names. For example, Alleppey Finger turmeric, Cochin ginger, Malabar pepper, and Sannam chilies have individual specifications differentiating them from other turmeric, ginger, pepper, and chilies, respectively. Each specification states limits for moisture, volatile oil, total ash, acid-insoluble ash, and starch in addition to the standards for extraneous matter, necessitating inspection of spices for each chemical/physical quality before export. For example, black pepper, one of the most important import items, is classified into more than 10 grades, depending upon the proportion of light berries, harvest place (Malabar or others), moisture content, and so on. Tellicherry black pepper in particular is classified by size. Curry powder, a mixture of spices, is graded according to the amount of spice or salt contained. Curry powder containing 85% or more and less than 5% salt is graded as "standard," and one with 70% or more and 10% or less salt is graded as "general."

3. Grade specifications for Sarawak pepper in Malaysia

These specifications, introduced by the Pepper Marketing Board, is designated for Sarawak pepper, which accounts for more than 90% of the total pepper production of Malaysia [6]. The grade of black pepper is determined according to the amount of light berries present, extraneous matter, moisture, and other characteristics. Standard Malaysian Black Pepper No. 1 (brown label) has the highest grade, followed by Sarawak Special Black (yellow label), Sarawak FAQ Black (black label), Sarawak Field Black (purple label), and Sarawak Coarse Field (gray label) with the lowest grade. There are also standards for white pepper, in which the amount of light berries, moisture, extraneous matter, and black pepper present are limited. White pepper is graded as follows: Standard Malaysian White Pepper No. 1 is highest (cream label), followed by Sarawak Special White (green label), Sarawak FAQ White (blue label), Sarawak Field White (orange label), and Sarawak Coarse White (gray label). In general, higher grade black/white pepper contains less moisture and fewer light berries as well as less extraneous matter.

4. Grading of nutmeg in Grenada and Indonesia

These specifications set limits not for export purposes, but for grading nutmeg of two major origins: Indonesia and Grenada. Nutmeg can be classified largely into "sound Nutmeg," which has sustained no injuries, and "substandard Nutmeg."

Sound Nutmeg is also graded as "80s" and "110s" according to the number of nutmeg per pound, for example, "80s" means there are 80 pieces contained in one pound. Substandard Nutmeg, which is exported from Indonesia, can be shriveled and "BWP" (broken, wormy, punky). In Grenada, broken or injured nutmeg is graded "defective." BWP and defective nutmeg are often used as raw material for ground nutmeg. But it is thought that broken or injured nutmeg is more likely to suffer from insect infestation, which sometimes causes microbial problems such as aflatoxin production.

5. Specification of paprika in Hungary and Spain

Spain and Hungary are among the major nations exporting paprika since the early 20th century. Specifications for paprika in Spain define paprika as the product obtained by dehydrating and then grinding clean, fully ripe berries of *Capsicum annum* and *Capsicum longum* and prohibit both the sale and the use of biologically altered paprika [4]. In Spain, paprika is classified into three grades according to moisture content, total ash, ether-soluble extract, acid-insoluble ash, and total fiber. Extra grade paprika is produced only from the peel (all seeds and placenta removed), Select grade allows 10% seed content, and Ordinary grade allows a 30% seed content.

In Hungary, grade and quality standards are specified by The Hungarian Office of Standard [4]. Paprika is classified according to three qualities and eight grades according to appearance, pungency, and other characteristics such as total ash and amount of ether extract. First-quality grades are Special Paprika, Table Quality Mild Paprika (nonpungent), Table Quality (mildly pungent), and "Hot" Table Paprika. Second quality grades include Semi-sweet Paprika, and third-quality grades include Pink(rose) Paprika and Pungent Paprika.

6. Specifications of other exporting nations

In addition to the above-mentioned countries, many spice-exporting nations have instituted standards and quality grades. Sri Lanka regulates the quality of cardamom, coriander, and other spices and the testing methods used under the Bureau of Sri Lanka Standards. The minimum volatile oil content is relatively high (4%), and six grades are identified according to appearance and color. Zanzibar and Madagascar grade cloves for export according to appearance, moisture content, extraneous matter, and other factors.

II. SPICE QUALITY

A. Insect Infestation

1. Harmful insects

Insects harmful to farm products, including spices, are usually controlled by agricultural chemicals during cultivation. But spices can also be damaged by insects, including mites, during storage because spices, like flour and beans, are generally stored for relatively long periods of time. Such pests are called "stored grain insects." Of the many harmful insects, moths and beetles are most damaging to spices. How fast the insects develop and breed depends on the atmospheric temperature, the kind of spice, as well as the kind of insect. Red pepper and basil are among the spices that often suffer from harmful insects during storage; parsley, garlic, and oregano do not. The cigarette beetle and Indian meal moth are typical problem insects found on spices. The cigarette beetle (*Lasioderma serricorne* Fabricius) is found in many areas from tropical to temperate zones. The Indian meal moth (*Plodia interpunctella* Hubner) is one of the most common insects attacking flour and rice. Besides these insects, the coffee bean weevil is known to breed on nutmeg [7]. Seenappa et al. [8] observed that both the flour beetle (*Tribolium confusum* Jacquelin Duval) and the red flour beetle (*Tribolium castaneum* Herbst) also bred on red pepper. So-called book lice (*Liposcelidae bostrychophilus* Badonnel) is known to breed on paper and sometimes can be transferred to spices from corrugated cartons or Kraft bags during transport. Miyazima et al. observed the attraction of Tyroglypid mites to several spices [9]. In this study, red pepper, coriander, garlic, laurel, and mandarin were put in trap vessels and the number of mites attracted to each vessel was counted. As shown in Table 2.1, red pepper and laurel were found to attract the mites the most after both 2 and 48 hours, while garlic did not attract them nearly as much.

TABLE 2.1 Attraction of Tyroglypid Mites to Certain Spices

	A. After 2 hours				B. After 48 hours	
	Number				Number	
Spice	1	2	3	Spice	1	2
Red pepper	74	55	45	Red pepper	469	156
	20.00%	18.27%	45.45%		43.66%	22.41%
Coriander	25	23	13	Coriander	220	257
	6.75%	7.64%	13.13%		20.48%	36.92%
Garlic	99	42	7	Garlic	87	66
	26.75%	13.95%	7.07%		8.10%	9.48%
Laurel	152	134	30	Laurel	247	123
	41.08%	44.51%	30.30%		22.99%	17.67%
Mandarin	20	47	4	Mandarin	51	94
	5.40%	15.61%	4.04%		4.75%	13.50%
Total	370	301	99	Total	1074	696
	99.98%	99.98%	99.00%		99.98%	99.98%

Source: Ref. 9.

2. Fumigation for insects

Insects found on spices breed and multiply very quickly, resulting in big problems unless appropriate measures are taken in the early stages. For example, Indian meal moths lay approximately 200 eggs at one time, and the life cycle of this moth is only around 40 days. Therefore, it can be expected that the total number would increase explosively within a short period of time. The most common means used to control insects in the warehouse is fumigation. The advantage of using fumigation is that it can reach every part of the storage warehouse and act uniformly. The chemicals most widely used on spices for insect disinfection purposes are methyl bromide and phosphine.

Methyl Bromide The boiling point of methyl bromide is 3.6°C; it can be used even in winter as a fumigant. The efficacy of this fumigant can be generally described by the equation:

$$K = C \times T$$

where K is the fumigation efficacy, C the gas concentration, and T the fumigation time. The efficacy of the fumigant is enhanced by a longer fumigation time or higher gas concentration. As for fumigation temperature, efficacy tends to increase as the temperature increases. The disadvantage of this fumigant is that it is not always as effective as phosphine, especially for pupae and eggs of some insects, in spite of its strong efficacy against adult insects. However, it has been used in warehouse for spices and other agricultural products for almost 50 years, so that relatively predictable fumigation effects can be expected. There are also some advantages to using methyl bromide: its fumigation time is relatively short (several hours to a couple of days) and it is relatively harmless to humans. For these reasons it is used as fumigant for many farm products, including spices. However, it is believed to react with ozone to deplete the ozone layer, and there are some groups that would restrict both the use and the production of this fumigant in order to protect the environment.

In November 1992, methyl bromide was listed as an ozone-depleting substance at the Montreal Protocol held at Copenhagen. With this in mind, the U.S. Environmental Protection Agency has decided that the use of methyl bromide will be banned in a few years [10], and other industrial countries plan to phase out its use afterward.

Phosphine (Aluminum Phosphate) Aluminum phosphate is usually produced in tablet form. It decomposes upon reacting with water in the air into hydrogen phosphate, or phosphine, which has a strong fumigation effect [11].

$$AlP + 3H_2O \rightarrow Al(OH)_3 + PH_3$$

Since phosphine is generated from aluminum phosphate very slowly, the time required for fumigation is relatively long compared to that with other fumigants. Although hydrogen phosphate is said to be more toxic to humans than methyl bromide, it has some advantages in that it has very strong fumigating effect on many sorts and forms of insects, including egg and pupae, and it is easy to deal with because it is a solid. As mentioned, the use of methyl bromide will be restricted in the future, and the use of aluminum phosphate is expected to increase concomitantly.

3. Other ways of exterminating insects

Most insects will die in atmospheres containing less than 2% oxygen [12]. Controlled-atmosphere storage in which carbon dioxide or nitrogen is substituted for oxygen in air has been studied and is practiced in some nations. Carbon dioxide has a particularly lethal effect against some insects and can be expected to be used more widely. Also, most insects from tropical regions cannot survive at 15°C or lower (except for mites, which multiply even at 10°C), and low-temperature storage is considered to be an effective method of eradicating them.

B. Microorganisms

1. Types of microbes

Most spices are cultivated and harvested in tropical and subtropical regions. Unfortunately, dehydration, separation, milling, and storage are not always conducted hygienically. Relatively high microbial counts on spice can be caused by contamination during postharvest processing as well as while the spice plant is growing. In most cases, harvested spice plants are spread on the ground and sun-dried for several days until the desired moisture level is reached. A large number of microbes, predominantly spore-forming bacteria (e.g., *Bacillus* spp.), exist in the soil at levels of around 10^6-10^8 cells/g. These can be not only a direct cause of contamination during the sun-drying process, but can also account for the relatively high counts of microbes on ginger and turmeric, which are dried rhizomes of tropical origin.

Examples of the mean counts of mold and bacteria for major spices that are not sterilized are shown in Table 2.2. As can be seen from this table, the number of bacteria existing in spices is generally higher than that found in molds. The principal bacteria and molds found depend on the kind of spice, the place of harvest, and the place of postharvest processing, but in general spore-forming bacteria like *Bacillus* spp. are predominant; *Staphylococcus* and *Streptococcus* spp. are also found in large numbers in most kinds of spices. However, it is unusual for a spice to be infected by pathogenic bacteria such as *Escherichia coli* or *Salmonella*. Individually, in case of clove, there are only small mold counts in number, and the total counts of microbes found on this spice are also relatively low. This tendency of lower counts of microbes on clove was supported by a

TABLE 2.2 Numbers of Microorganisms in Ground Spices

	Propagules/g spice		
Spice	Actinomycetes	Bacteria	Molds
Aniseed	0	1.9×10^5	9.5×10^3
Black pepper	0	5.2×10^7	6.4×10^5
Cardamom	2.5×10^3	7.6×10^6	1.6×10^3
Caraway	0	5.2×10^4	1.5×10^3
Red pepper (cayenne)	0	2.2×10^6	3.9×10^4
Celery	0	2.6×10^6	1.2×10^3
Cinnamon	0	3.5×10^5	8.7×10^4
Cloves	6.0×10^2	4.3×10^4	0
Coriander	0	3.7×10^6	1.3×10^5
Cumin	0	2.1×10^5	1.5×10^3
Fennel	2.5×10^3	1.6×10^5	6.7×10^3
Fenugreek	6.5×10^2	6.2×10^4	2.5×10^3
Ginger	1.0×10^2	8.7×10^6	1.7×10^3
Jamaica pepper (allspice)	0	6.8×10^6	7.0×10^4
Mace	0	6.4×10^4	8.0×10^2
Mustard	0	5.8×10^6	2.7×10^3
Nutmeg	4.0×10^3	1.1×10^5	6.2×10^4
Paprika	0	6.6×10^6	5.5×10^2
Turmeric	0	4.8×10^6	0.5×10^2
White pepper	0	6.8×10^4	6.5×10^4
Mixtures			
Curry powders	0	2.0×10^5	9.2×10^4
Garam masala	0	2.4×10^6	4.1×10^4
Mixed baking spice	0	6.8×10^4	3.2×10^4

Source: Ref. 13.

couple of past research [13,14]. As we explain in "antimicrobial activity of spice" in later chapter, eugenol in essential oil of clove has strong antimicrobial activity, and it is therefore considered that low counts of microbes on this spice are due to its essential oil which seeps out into the surface of ground clove.

2. Molds and aflatoxin

As mentioned above, almost no pathogenic bacteria exist in spices. It is, however, necessary to pay attention to some types of molds that can produce toxic substances. Typical examples are *Aspergillus flavus* and *Aspergillus ochraceus*, which produce hazardous toxins called aflatoxin and ochratoxin, respectively.

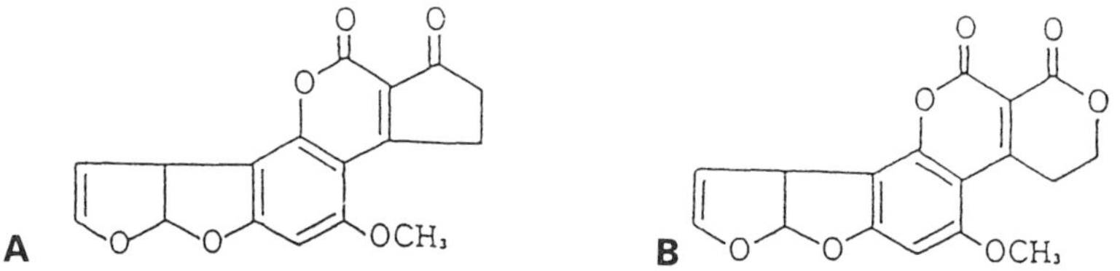

FIGURE 2.1 Chemical structure of (A) aflatoxin B1 and (B) aflatoxin G1.

Aflatoxin is especially known to be a cause of cancer and is classified into B1, B2, G1, and G2 types (Fig. 2.1). Many European nations also limit aflatoxin in spices to no more than 10 ppb. According to many past investigations and inspections, the types of spices likely to exceed this limit are limited to red pepper and nutmeg. In Japan, aflatoxin was detected in peanut butter for the first time in 1970, and the Ministry of Welfare the following year set an upper limit for aflatoxin B1—the most toxic among all aflatoxin types—of 10 ppb, and decided to obligate importers to check aflatoxin levels of peanuts and peanut products. The Ministry of Welfare made known its intention to strengthen the supervision of aflatoxin contamination in nutmeg in 1985 and in red pepper somewhat later, based on inspections conducted on many import foods.

Aspergillus glaucus and *Aspergillus niger* are the predominant molds found to contaminate spices. *Aspergillus flavus*, which includes strains capable of producing aflatoxin, has also been confirmed on many kinds of spice, although in most cases *A. flavus* does not outnumber *A. niger* or *A. glaucus*. In addition to *Aspergillus* spp., a large number of *Penicillium* spp. have been detected on fenugreek, ginger, and anise seed [13].

Molds detected on spices are generally called "storage fungi" because they usually adhere to and increase their number on spices while being sun-dried or stored after harvesting. It is also known that some types of beetles convey molds, including *Aspergillus* spp., to some types of spices, such as red pepper. It is, therefore, important not only to control the relative humidity of the storage warehouse, but also to exterminate insects by fumigation in order to prevent mold colonization and aflatoxin production during storage. Seenappa et al. observed the influence of relative humidity on the colonization of *Aspergillus* spp. for different parts of red pepper [15]. According to this research, *A. niger* was the predominant mold found on stalks during storage at 28°C and 85% relative humidity, but when the relative humidity was raised to 95%, *A. flavus* and *A. ochraceus* predominated on the stalks and the pods and the possibility of aflatoxin production was enhanced. Internal colonization by molds occurred in mechanically damaged pods. As molds grow inside the pods, they gradually become a yellowish color and the color of the seeds inside the pods changes from the normal ivory white to olive green in pods containing *A. flavus* and to black in pods containing *A. niger*. This means that red peppers heavily contaminated by *Aspergillus* spp. can be easily identified by their yellow discoloration, and removal of all discolored pods after storage would be a useful means of preventing aflatoxin-contaminated red pepper.

C. Sterilizing or Pasteurizing Methods

Several different means of sterilization have been examined to decrease the number of microbes on spices. The three methods discussed here are the ethylene oxide gas method, irradiation, and steam sterilization.

1. Ethylene oxide gas method

Ethylene oxide has been widely used in a cold sterilization method, especially for spices, because it has a minimal influence on the very susceptible flavors and aromas contained in spices. Ethylene oxide has a strong sterilizing and fumigating effect because it reacts with active reactants of protein, including carboxyl groups, sulfhydryl groups, amino groups, and hydroxyl groups, to change the protein structures in living insects and microbes. Several reports show its sterilizing effects on different spices and sterilizing effects were confirmed for many of the spices tested, such as black pepper and paprika.

Koizumi et al. [16] studied the influence of temperature on sterilization of black pepper with ethylene oxide and found that a temperature range of 25–30°C is most effective for sterilizing microbes on this spice. The sterilizing efficiency is decreased drastically at temperatures of less than 10°C.

Despite its strong sterilizing effect, ethylene oxide is known to be toxic to humans in the vapor concentration range used when sterilizing [16]. Because of its toxic nature, there are many nations which restrict its use in foods. Ethylene oxide does not stay in spices for long periods of storage because the boiling point of ethylene oxide is around 10°C, as shown in Table 2.3 [18]. Usually, ethylene oxide residue is decreased to less than one-tenth after a week of storage. There are some reports that it sometimes reacts with nutrients to change their natures because its reactability is very high [19,20]. In particular, ascorbic acid (vitamin C) and riboflavin (vitamin B_1) were reported to be decomposed in reaction with ethylene oxide [19]. Some reports showed that ethylene oxide treatment was not always desirable for some spices, such as paprika, which may turn a dark, dull color [20]. Table 2.4 shows the comparative effects of ethylene oxide and gamma-irradiation on the color of paprika.

TABLE 2.3 Boiling Points of Ethylene Oxide, Ethylene Chlorohydrin, and Ethylene Glycol

	Boiling point (°C)
Ethylene oxide	10.7
Ethylene chlorohydrin	128.8
Ethylene glycol	197.6

2. Irradiation

Radiation types used for food irradiation are currently limited to gamma-radiation generated from Cobalt-60 and Cesium-137 and electron and X-radiation from an electron accelerator.

In 1980, the Joint Expert Committee on the Wholesomeness of Irradiated Food (JECFI) concluded that "irradiation of any food commodity up to an overall average dose of 10 kGy introduces no toxicological hazard," and accordingly in 1983 a Codex General Standard for Irradiated Foods and a Recommended International Code of Practice for Operation of Radiation Facilities Used for Treatment of Food were adopted [17]. To date, 37 nations permit the use of food irradiation for 40 or more kinds of foods, with the foods for which irradiation is permitted varying according to country. As for spices, 32 governments have regulations allowing the use of irradiation for spices, but in Japan spice irradiation is still banned.

TABLE 2.4 Comparative Effect of Ethylene Oxide and Gamma-Irradiation on Color of Paprika

Treatment	Color of ground paprika			Water-soluble color (O.D. at 450 nm)	Water-insoluble color (ASTA units)
	(brightness) L	(redness) + a	(yellowness) + b		
Raw paprika	24.8	22.7	12.2	0.23	85.7
Ethylene oxide–treated paprika	22.26	16.3	9.7	0.33	81.0
Gamma-irradiated paprika	24.9	23.6	12.3	0.23	85.7

Source: Ref. 20.

Spice Qualities and Specifications

Several reports exist regarding the sterilizing effects of irradiation and its effect on the qualities and constituents of spices. Vajdi and Pereira [20], in a study of gamma-irradiation treatment of six different spices, found that thermophilic bacteria existing on spices with a population of 10^3-10^5 cells/g and aerobic spores with populations of 10^2-10^5 cells/g were sterilized by irradiation doses of 10 kGy and 4 kGy, respectively (Table 2.5). Eiss [21] reported average doses of 6.5–10 kGy were enough to sterilize coliform bacteria and fungi existing on spices and to reduce standard plate counts to less than 3000 cells/g. The effect of irradiation treatment on the qualities of 12 different spices was reported by Bachman and Gieszczynska [22]. They found that spice flavors remained unchanged below a radiation dose of 10 kGy, except for coriander and cardamom, which changed in flavor at a dose of 7.5 kGy (Table 2.6). Quantitative changes of constituents contained in the essential oil of black pepper and other spices after gamma-irradiation have been examined by some researchers. Of a variety of volatile compounds in black pepper, α-pinene and carene were reported to be irradiation-resistant properties, while terpene compounds such as α-terpinene, γ-terpinene, and terpinolene tended to decrease in quantity even at a dose of 7.5 kGy [23]. Peroxide values for extracted lipids from some irradiated spices, especially nutmeg, were found to be increased when relatively high doses were applied [24]. The changes in peroxide value for several spices due are shown in Figure 2.2. The increases were slight, however, when the dose was below 10 kGy. It is clear from these examples that negligible irradiation doses of between 7 and 10 kGy can reduce the populations of microbes on spices without changing their essential qualities.

3. Steam sterilization

Heat sterilization is the most popular sterilizing method for liquid foods, such as sauces, but for low-moisture products, such as spices, this method is inadequate because the thermal conductivity inside such dried products is very low. In practice, when *Clostridium sporogenes*, a thermophilic spore bacteria, is to be sterilized by heat, a D-value (the time required to reduce the population of microbes by one digit) at 120°C with wet heat is 0.14–1.4 minutes, while that with dry heat is 115–195 minutes [25]. It is for this reason that steam sterilization with saturated or supersaturated steam is considered to be most effective for the purpose of reducing microbes on spices. Kikkoman has produced an "air stream sterilizing system" for which superheated steam is used as the heating medium. Superheated steam is a kind of dry steam, which does not cause the target dry ingredient to become wet or to stick together during sterilization. In this system, the dry ingredient is delivered to the air stream tube, where superheated steam sterilizes it while it is being transported. One of the advantages of this system is that even thermophilic spore bacteria can be sterilized quickly. The effects of

TABLE 2.5 Comparative Effect of Ethylene Oxide and Gamma Irradiation on Bacteria Flora of Selected Spices

	Treatments (no. of organisms)								
	Raw			Ethylene oxide			Gamma-irradiation		
Spices	Total count	Thermo-philic	Aerobic spores	Total count	Thermo-philic	Aerobic spores	Total count	Thermo-philic	Aerobic spores
Black pepper	4.0×10^6	1.58×10^6	6.34×10^4	1.48×10^3	4.3×10^2	0.0	0.0	0.0	0.0
Paprika	9.86×10^6	3.24×10^5	3.0×10^3	0.0	0.0	0.0	0.0	0.0	0.0
Oregano	3.26×10^4	1.8×10^3	1.0×10^2	0.0	0.0	0.0	0.0	0.0	0.0
Allspice	1.74×10^6	1.5×10^6	1.05×10^2	4.25×10	0.0	0.0	0.0	0.0	0.0
Celery seed	3.7×10^5	1.3×10^5	3.94×10^3	0.8×10	0.0	0.0	0.0	0.0	0.0
Garlic	4.65×10^4	9.0×10^2	0.0	1.45×10^4	3.5×10^2	0.0	0.0	0.0	0.0

Source: Ref. 20.

Spice Qualities and Specifications

TABLE 2.6 Radiation Dose Required to Produce a Change in Flavor of Certain Spices

Spice	Dose (kGy)	Spice	Dose (kGy)
Caraway	12.5	Marjoram	7.5–12.5
Coriander	7.5	Black pepper	12.5
Cardamom	7.5	Pepper substitute	12.5
Charlock	10.0	Pimento	15.0
Juniper	>15.0		

Source: Ref. 22.

steam sterilization on microbes and other spice qualities are shown in Table 2.7 [26]. Although the essential oils of some spices decreased in content, total microbe counts were found to decrease. Although the sterilization time with this system is very short, heat permeation into the target ingredient is not as high as with ethylene oxide or radiation. Therefore, the sterilizing effect may vary widely within a lot, especially when the surface of spice is not smooth, as in the case of whole seed. Other disadvantages of this system are that green-colored spices, such as basil and parsley, tend to become brown with high temperature.

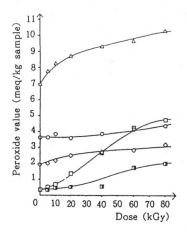

FIGURE 2.2 Changes in peroxide value in spices after gamma-irradiation. (△), Rosemary; (○), black pepper; (◊), clove; (□), nutmeg powder; (■), whole nutmeg. (From Ref. 24.)

TABLE 2.7 Sterilizing Effect of Superheated Steam Sterilization on Spices

Spice	Sterilizing condition (kg/cm$_2$ G-C°-sec)		Moisture (%)	SPC (cells/g)	Coliform	Other qualities
Black pepper, whole	1.	No sterilizing	11.7	3.5×10^7	+	3.0% essential oil
	2.	3-180-7	11.0	<300	−	3.0% essential oil
Paprika, ground	1.	No sterilizing	10.9	1.6×10^6	−	9.49 mg% carotene content
	2.	1-135-7	11.2	<300	−	9.04 mg% carotene content
Turmeric, ground	1.	No sterilizing	10.1	8.1×10^6	+	3.1% curcumin content
	2.	1.5-140-5	7.6	7.3×10^2	−	2.9% curcumin content
Bay leaves, ground	1.	No sterilizing		6.8×10^3	−	2.5% essential oil
	2.	0.2-120-3		<300	−	2.0% essential oil

Source: Ref. 26.

D. Chemical Properties

The function of a spice is to give foods a typical flavor or color. In addition to water content specifications for the purpose of preventing mold growth, chemical standards for quality control are also necessary. The official specifications for importing and exporting nations do not always set such standards, but most spice-processing companies do this themselves in order to provide the consumer with high-quality products. Some of these chemical standards are explained below.

1. Pungency standards

Black and White Pepper The major pungent compound in both black and white pepper is an alkaloid compound called piperine. Besides this component, pepper contains small amounts of other alkaloid compounds, such as piperittine and piperyline, but piperine accounts for around 98% of the total alkaloid compounds contained in these spices. Piperine content of both black and white pepper tends to be influenced by the many environmental factors, such as climate, temperature, and especially where it was grown (harvest place). For example, Brazilian black pepper tends to contain a relatively low amount of piperine compared to black pepper of other origins. The chemical structures of piperine, piperittine, and piperyline are shown in Figure 2.3.

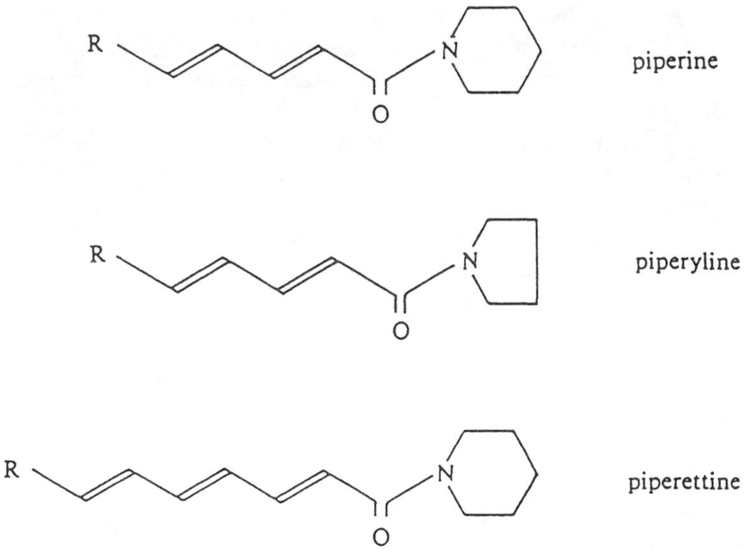

FIGURE 2.3 Chemical structures of piperine, piperyline, and piperettine. (From Ref. 4.)

Quantitative analysis for piperine includes the ultraviolet (UV) spectrophotometric method (used by the ASTA), the gas-chromatographic (GC) method, infrared analysis, the Kjeldahl method, and some calorimetric methods, among others. The Kjeldahl method, which measures the nitrogen levels in piperine, is a simple way to analyze this pungent compound, but the test results tend to be high with this method because it includes chavicine, piperittine, and some amino acids contained in black and white pepper in its measurements. Colorimetric methods measure the production of formaldehyde, which is produced by reacting chromotropic acid with the CH_2O_2 of piperine in the presence of concentrated sulfuric acid [27]. The UV spectrophotometric method, based on absorption in the UV region at 343 μm is used most frequently because, besides being both simple and rapid, it has a high specificity for piperine and does not tend to capture other compounds such as chavicine and piperine isomers [28]. Furthermore, it can also determine piperettine content at a different ultraviolet region. The GC method utilizes a reaction that releases formaldehydes. It, therefore, analyzes not only piperine but also other compounds such as dextrose. Recently, the high-performance liquid chromatography (HPLC) has been studied, and the analytical results are said to be a little lower than with the UV method because its resolution is superior.

Red Pepper The pungent principles of red pepper are capsaicinoid compounds, which are vanillylamides of monocarboxylic acids. The chain length of each capsaicinoid has a drastic effect on the degree of pungency [29]. The major pungent compound contained in this spice is capsaicin, which accounts for over 50% of the total capsaicinoids, dihydrocapsaicin, norhydrocapsaicin, homodihydrocapsaicin and homocapsaicin are also contained in relatively high amounts. The chemical structure of capsaicin is shown in Figure 2.4. The degree of pungency is known to differ widely depending not only on species and harvest place, but also on the growing and drying conditions and harvest year. An almost twofold difference in pungency levels was reported, depending upon geographical location [29,30]. It is also known that maturity of the plant significantly influences pungency level. For the reasons described, red pepper pungency is usually checked quantitatively by spice-processing companies before they use it.

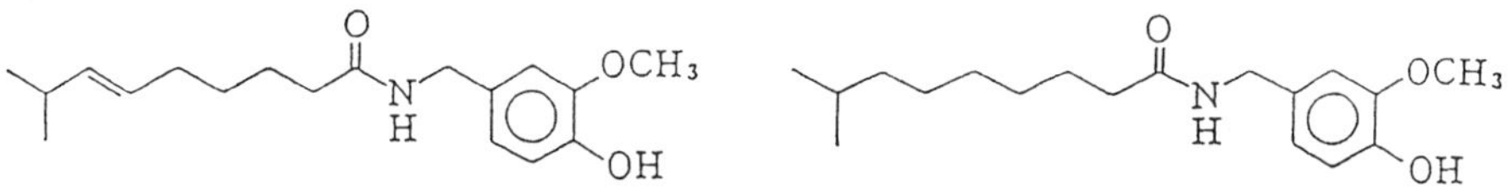

capsaicin   dihydrocapsaicin

FIGURE 2.4 Chemical structure of capsaicin.

To measure the degree of pungency, the Scoville heat value method, which measures pungency by determining a distinct pungent sensation and was developed early in this century, is still widely used. It has been adopted as an official ASTA method [31] because of its simplicity in terms of analytical apparatus and theory. But this method has several drawbacks, including the long period of time required to train panels and the difficulty in maintaining accuracy because of panel fatigue. In addition to this organoleptic method, numerous colorimetric methods has been used extensively. More than 10 kinds of pungent compound have been identified in red pepper, with a different level of pungency for each. It is, therefore, difficult to determine the overall pungency of the pepper as well as to analyze each pungent compound quantitatively. One colorimetric method measures the blue color generated by the interaction of a vanadium oxytrichloride with the hydroxyl group of capsaicin or capsaicin isomers, based on the molybdenum blue reaction [29]. Several ultraviolet methods have also been reported. One UV method was compared with the Scoville method, and a very close correlation between these two methods was found [30]. Mori et al. also found adequate correlation between a modified UV method and the Scoville heat value [32]. Many methods based on gas chromatography have been also developed. Hollo et al. [33] compared their gas chromatographic procedure with a colorimetric method, and found that it produced a lower value than the colorimetric method. They posited that the capsaicin analog might be also measured by the colorimetric method. HPLC methods for capsaicin analysis have also been studied, and one has been adopted as an ASTA official method. As with black pepper, the HPLC method is considered to be more accurate because of its high resolution, especially for capsaicin analysis in curry powder or curry roux, which contains impurities. However, all these quantitative methods have the disadvantage of taking a long time because they are all based on the extraction process, and simpler and easier direct analytical methods are expected to be developed.

Mustard The pungent principle of this spice is a variety of isothiocyanate compounds, which exists in plant tissue as a glycoside. The major pungent compound of black mustard is allyl isothiocyanate, which is produced by the reaction of the enzyme myrosinase with the glycoside sinigrin, while that of white mustard is *p*-hydroxybenzyl isothiocynate, which is less pungent than allyl isothiocyanate. The *p*-hydroxybenzyl isothiocyanate is generated by the action of myrosinase on the glycoside sinalbin. Because these pungent compounds exist in the tissue in a form of glycosides, which do not have any pungency, pungent compound levels usually do not decrease even during manufacturing processes such as drying or milling. For this reason the pungency level of mustard does not vary widely. The chemical structures of allyl isothiocyanate and *p*-hydroxybenzyl isothiocyanate are shown in Figure 2.5.

Both allyl isothiocyanate and *p*-hydroxybenzyl isothiocyanate can be analyzed quantitatively with gas chromatography, which can analyze other isothio-

R : $CH_2=CH-CH_2-N=C=S$

Allyl isothiocyanate

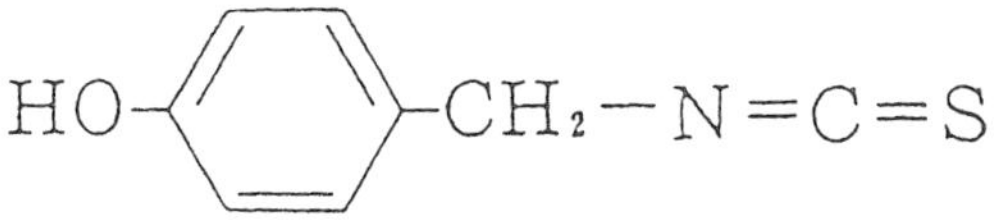

$HO-\langle\bigcirc\rangle-CH_2-N=C=S$

Benzyl isothiocyanate

FIGURE 2.5 Chemical structures of allyl isothiocyanate and benzyl isothiocyanate.

cyanate compounds as well. The colorimetric method for allyl isothiocyanate is used by the ASTA. But with this colorimetric method, p-hydroxybenzyl isothiocyanate and some other isothiocyanate compounds cannot be analyzed quantitatively. Because these pungent compounds are considered to be stable during storage, pungency levels and allyl isothiocyanate content are not always measured quantitatively as a routine quality control activity at spice companies.

Other Spices Pungency is an important property in other spices, although their pungency levels are not routinely checked quantitatively. Sansho, or Japanese pepper, is characterized by its "numbing" pungency due to both α-sanshool and β-sanshool. The chemical structure of this compounds is shown in Figure 2.6. Since both compounds degrade quickly after the pepper is ground, it is usually stored in its whole form.

The major chemical compounds that account for the typical pungency of ginger are gingerol, zingerone, and shogaol. Gingerol, the most pungent of the three, is thought to convert to shogaol due to dehydration of the β-hydroxyl ketone during the drying process, and shogaol is degraded into nonpungent polymers

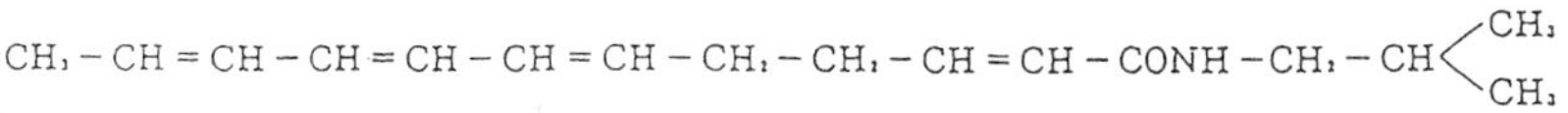

$CH_3-CH=CH-CH=CH-CH=CH-CH_2-CH_2-CH=CH-CONH-CH_2-CH\langle^{CH_3}_{CH_3}$

Sanshool

FIGURE 2.6 Chemical structure of sanshool.

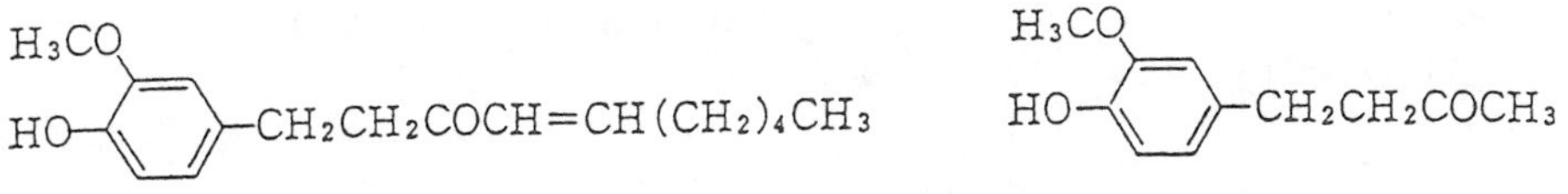

Spice Qualities and Specifications

$$H_3CO\text{-}C_6H_3(OH)\text{-}CH_2CH_2COCH=CH(CH_2)_4CH_3$$

shogaol

$$H_3CO\text{-}C_6H_3(OH)\text{-}CH_2CH_2COCH_3$$

zingerone

FIGURE 2.7 Chemical structures of shogaol and zingerone.

under the influence of heat. Figure 2.7 shows the chemical structures of zingerone and shogaol. The pungency of ginger varies depending on the drying conditions.

2. Flavor/aroma

Spice flavors are expressed by their volatile essential oils, and some official import/export specifications set lower limits for volatile oil content for some spices. Since flavor is the most important property of most spices, many spice companies set their own standards without official specification or set higher standards than do official specifications. Because volatile oils tend to decrease in amount during processes like grinding or sterilization, minimum amounts in ground spices are usually lower than in whole spices. The approximate volatile oil content of each major spice is listed in Table 2.8. The analytical procedure for

TABLE 2.8 Approximate Amounts of Essential Oil in Certain Spices

Spice	Essential oil (%)	Spice	Essential oil (%)
Allspice	3.0–5.0	Sansho	2.0–6.0
Anise seed	2.0–3.0	Marjoram	0.7–3.5
Basil	0.2–0.4	Nutmeg/Mace	7.0–16.0
Bay leaves	0.7–7.0	Onion	0.02–0.06
Caraway	3.0–6.0	Oregano	0.2–0.4
Cardamom	3.0–8.0	Parsley	0.1–0.3
Cinnamon	1.0–2.4	Pepper	1.0–4.0
Clove	15.0–18.0	Rosemary	0.7–2.0
Coriander	2.0–3.0	Saffron	0.5–1.0
Cumin	2.4–4.0	Sage	1.5–2.5
Dill	1.0–4.0	Tarragon	0.2–0.8
Fennel	1.0–6.0	Thyme	0.7–2.5
Garlic	0.1–0.2	Turmeric	1.0–6.0
Ginger	0.4–4.0		

essential oil content is called the Clevenger method, which extracts the oil by distillation. In some cases the volatile oils in spice leaves (herbs) are not listed in specifications because their contents are generally much lower than those of seeds and tend to vary widely with the ASTA methods.

It is also possible to measure a specific compound in an essential oil to check the flavor characteristic of a certain spice. In the case of cinnamon, cinnamaldehyde, which accounts for 60–90% of the entire composition of its volatile oil and is considered to represent its flavor characteristic, is sometimes listed in specifications in addition to volatile oil content. The amount of cinnamaldehyde is determined by indirect titration for its steam volatile oil. Even within one species of spice, qualitative and quantitative differences in individual volatile composition exist depending on product origin or spice cultivator. Russell et al. [34] observed volatile differences between the Lampong and Sarawak cultivars of *Piper nigrum* (black pepper). Although these differences should affect the characteristics of spice flavors, few standards exist to control their delicate flavors.

3. Color

Parsley The green color of parsley is a very important property of this spice. This green color, expressed by chlorophylls, is usually checked using an organoleptical method because of the difficulty of quantitative measurement. Furthermore, the color tone of parsley can be judged by appearance, because it is not concentrated as is the color of paprika or turmeric.

Paprika Paprika is used primarily for its coloring properties, and its color tone is important in determining the value of this spice. Its typical yellowish-red color is formed by approximately 10 different kinds of carotenoids, with capsanthin predominating (>30% of the total carotenoids). The other carotenoids, which account for 0.1–0.8% of the total weight of this spice, include zeaxanthin, capsorbin, β-carotene, and cryptoxanthin. The color value is generally determined by measuring an acetone extract spectrophotometrically with an absorbance at 460 nm. This method is described in ASTA. Paprika loses its color during storage due to oxidation, catalyzed by light, of carotenoid compounds. Since oxidation accelerates with increasing temperature, some manufacturing firms keep dehydrated paprika in a cold place until it is packaged. Also, the antioxidant ethoxyquin is sometimes added to prevent rancidity and slow down carotenoid degradation. However, ethoxyquin use is not allowed in most European nations or in Japan.

Turmeric The color of turmeric is, like that of paprika, is important to its commercial value. The yellow pigment obtained from the rhizomes of this spice is curcumin, one of the diketones. Curcumin, accounting for about 3% of this spice, is not water soluble but is soluble in alcohol.

The curcumin content of turmeric differs depending upon the harvest region. For example, Alleppy turmeric in India is characterized by its high amount of curcumin, and export specifications for curcumin content of turmeric for this region are high (6.5% minimum). Various analytical methods have been used to determine the color value of this spice. Color value is generally determined spectrophotometrically with absorbance at 425 nm, which has been adopted as an official method of ASTA. Jentzsch et al. described another analytical method using thin-layer chromatography [35].

Saffron The commercial value of saffron is determined mainly by its typical flavor, but its yellowish-red color is also an important property. The principle of this color is a terpene glycoside called crocin, which is a water-soluble yellowish-red pigment. Crocin is hydrolyzed to the sugar gentiobiose and crocetin, which belongs to the carotenoids. The amount of crocin in saffron is determined by observing the color of its water extract.

4. Other quality standards

In addition to the above-mentioned quality standards, there are cases in which total acid, acid-insoluble ash, or water-insoluble ash are quantified, mainly for the purpose of limiting the amount of extraneous matter in spices. For example, Sansho, a ground pericarp fruit skin, will have a very high acid-insoluble ash value if too many seeds or other parts of the plant are mixed with the pericarp when processed.

REFERENCES

1. Preface, American Spice Trade Association.
2. U.S. Federal Specification: Spices, ground and whole, and spice blends, No. EE-S-631H, 1975.
3. T. Matsukura, *New Food Industry*, 19(3): 8 (1977).
4. *Spices*, Shokuhin Shuppan, Tokyo, 1985.
5. *Agmark grade specifications*, The Spices Export Promotion Council.
6. *Grade Specifications, Labels and Certifications for Sarawak Pepper*, Pepper Marketing Board, Malaysia.
7. N. Watanabe, *Yunyunousanbutsu no Boutyu, Kunzyo Handbook* (H. Nakakita et al., ed.), Science Forum, Tokyo, 1995, p. 32.
8. M. Seenappa, L. W. Stobbs, and A. G. Kempton, *Int. Biodeterior. Bull.*, 15(3): 96 (1979).
9. H. Miyazima, A. Koraku, and T. Yamazaki, *Shokuhin Eisei Gakkaishi*, 10: 92 (1969).
10. A. Tateya, *Shokubutu Keneki*, 47(4): 37 (1993).
11. Y. Soma, *Yunyunousanbutsu no Boutyu, Kunzyo Handbook* (H. Nakakita et al., ed.), Science Forum, Tokyo, 1995, p. 32.
12. H. Nakakita, *Shokubutsuboueki*, 40(7): 1 (1986).
13. B. Flannigan and S. C. Hui, *J. Appl. Bacteriol.*, 41: 411 (1976).
14. J. Yesair and O. B. Williams, *Food Res.*, 7: 188 (1942).

15. M. Seenappa, L. W. Stobbs, and A. G. Kempton, *Phytopathology*, 70: 218 (1980).
16. T. Koizumi, S. Miyamoto, I. Tomiyama, and A. Suzuki, *Shokuhin Eisei Gakkaishi*, 5(3): 215 (1964).
17. P. Loaharanu, *Food Technol.*, 48: 124 (1994).
18. Y. Hosogai, *Boukin Boubai*, 6: 28 (1978).
19. Y. Hosogai, *Shokuhin to Kagaku*, 10: 114 (1973).
20. M. Vajdi and R. R. Pereira, *J. Food Sci.*, 38: 893 (1973).
21. M. I. Eiss, *Food Technol. Australia*, 36(8): 362 (1984).
22. S. Bachman and J. Gieszczynska, *Aspects of the Introduction of Food Irradiation in Developing Countries*, IAEA, STI/PUB/362, 1973, p. 43.
23. K. Kawashima, *Nippon Shokuhin Kogyo Gakkaishi*, 28: 52 (1981).
24. N. Kaneko, H. Ito, and I. Ishigaki, *Nippon Shokuhin Kogyo Gakkaishi*, 38: 1025 (1991).
25. I. Shibazoki, *Shin-Shokuhin-Sakkinkougaku*, Korin, Tokyo, 1983, p. 14.
26. N. Tsukada, *Shokuhin Kogyo*, 3: 34 (1984).
27. K. Mori, Y. Yamamoto, and S. Komai, *Shokuhin Kogyo Gakkaishi*, 21: 466 (1974).
28. S. Pruthi, I. O. & S. J. (12): 167 (1970).
29. A. Mega, *Crit. Rev. Food Sci. Nutr.*, 6: 177 (1975).
30. J. I. Suzuki, F. Tausig, and R. E. Morse, *Food Technol.*, 11: 100 (1957).
31. *Official Analytical Method*, American Spice Trade Association: Method 21, 1968.
32. K. Mori, H. Sawada, and Y. Nishiura, *Nippon Shokuhin Kogyo Gakkaishi*, 23: 199 (1976).
33. J. Hollo, E. Kurucz, and J. Bodor, *Lebensm. Wiss. Technol.*, 2: 19 (1969).
34. G. F. Russell and J. Else, *J. AOAC*, 56: 344 (1973).
35. K. Jentzsch, P. Spiegal, and R. Kamitz, *Sci. Pharm.*, 36: 257 (1968).

3

Cooking with Spices

I. THE EFFECT OF COOKING ON SPICE FLAVORS AND FUNCTION

The four major effects of spices used in cooking are flavor, pungency, coloring, and deodorizing. In general, each spice fulfills one major function, but some spices fulfill multiple functions. The following factors affect spice function, including flavor, when spices are used for cooking.

1. Spices contain both volatile and nonvolatile oils. Most spice flavors are a result of components included in the volatile oil, but volatile flavor compounds can be generated even from nonvolatile oil when heated. In general, flavor components tend to become weak or disappear at high temperatures. The flavor components of spices are more volatile with increased temperature. The flavor sensation of each spice

can be changed by controlling cooking conditions including temperature and time.
2. Oily components including volatile oils are generally soluble in alcohol, and the flavor of a spice can be enhanced effectively by cooking with alcohol.
3. The oily component of any spice is of course oil soluble. Therefore, flavor characteristics of spices can be enhanced by adding the spice to the oil phase of a salad dressing or to the oil used in the cooking process.
4. It is not always necessary to ingest the spice itself to achieve its desired effect. A deodorizing or neutralizing effect can be achieved, for example, without using the spice flavor per se.
5. It is sometimes effective to put a spice directly on the fire when grilling, resulting in a smoking effect.
6. Each spice contains different pungent components with different pungent characteristics. Therefore, there are differences in pungent sensation and in how long each can be preserved and in how each appears. It is important to put each pungent spice to its proper use in cooking. Some spices undergo changes in pungency due to the heating. Spices whose pungent components function enzymatically are not suitable for cooking with heat.
7. Coloring compounds contained in spices are classified as oil-soluble or water-soluble. It is necessary to put each type of spice to its proper use to achieve the coloring effect desired.

TABLE 3.1 Factors That Influence Spice Flavor

	Factors
Suitability with spices	
Raw materials	Chicken, seafood, egg, milk, vegetable, grain
Seasonings	Sweeteners, salt, soy sauce
Base solution	Water, oil, alcohol
Change of essential oils	
Timing for adding	Before cooking, during cooking, after cooking
How added	Sprinkle over material, knead into material
Cooking method	Picking, boiling, steaming, frying, deep-frying, baking
Cooking temperature	$<100°C$, $101–150°C$, $151–200°C$, $>201°C$
Change of taste	
To be eaten	Will be eaten, will not be eaten
Degree of processing	Fresh, dry (dehydrated), processed
Form of spice	Whole, ground, liquid, other

8. The flavor of food can be changed by using the proper spice at the proper time. For example, when the deodorizing effect of a spice is desired it is more effective to use the spice at the preparation stage or the finishing stage rather than during cooking.
9. In most cases, it is best to use a blend of spices rather than only one kind. The total amount of spice to be used in this case is greater than if only one spice is used because typical flavor of each spice is in general weakened when blending.
10. The flavor of a spice blend will be milder after a period of time than if it were used immediately after blending.
11. There are synergistic effects and suppressive effects for some spices. In other words, flavor characteristics, including flavor strength, are changed when it is used with another spice or seasoning.

Cooking factors that influence the flavor and some functions of spices are summarized in Table 3.1.

II. SPICE FORM AND FLAVOR

Spices can be used either fresh or in a dried form. A spice is in general characterized by its original fresh flavor, but some spices' grassy smell make them undesirable in the fresh form. Drying or milling processes may also change the original flavor of a spice or may activate an enzyme activity that results in an undesirable flavor.

When a spice is used for cooking, it is necessary to evaluate its overall quality, including microbial status, dispersion, heat-resistance, and color tone achieved in addition to its flavor. The different spice forms are summarized in Figure 3.1 [1]. Table 3.2 compares the different forms of spice based on suitability for use. Mashing a fresh spice results in a spice paste. Spice which is dried is called whole spice, and grinding this produces ground spice. These "half-processed" spices can be processed further to be suitable for various food-processing uses. For example, ground spice is further processed to yield a natural essential oil by steam distillation, or to yield an extracted spice such as oleoresin using alcohol or acetone. Such processed spices have the advantage of being uniform in quality and can be further processed to make other types of processed spice.

Liquid spice is manufactured by dissolving a spice oleoresin or essential oil in vegetable oil or alcohol. Emulsified liquid spice is made by emulsifying essential oil or oleoresin in water with emulsifier and stabilizer. But emulsified liquid spice is not always stable due to the separation of the oil phase from the water phase and to the recrystallization of spice components. Absorbent spice is produced by absorption of a spice oleoresin or essential oil to glucose, dextrin, or salt. But sometimes the spice component is easily volatilized or deteriorated depending upon the kind of absorbent used. Coating spices can be divided into

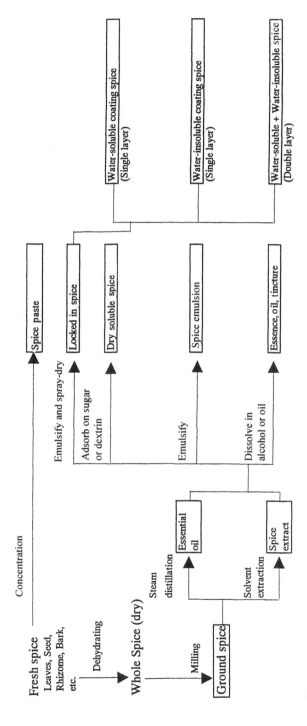

FIGURE 3.1 Different forms of spice.

TABLE 3.2 Comparison of Spice Characteristics

Property	Natural spice		Spice extract					Cooked in spice	
	Fresh	Dry	Spice paste	Essential oil tincture	Spice emulsion	Dry soluble spice	Water-soluble (single layer)	Water-insoluble (single layer)	Water-soluble–water-insoluble (double layer)
Flavor	⊙	○	△	△	△	△	△	△	□
Dispersiveness									
Water	×	△	○	○	⊙	△	⊙	×	⊙
Oil	△	○	⊙	⊙	⊙	⊙	⊙	○	⊙
Durability									
Color	×	△	□	△	△	△	○	○	⊙
Flavor	×	△	□	□	○	○	○	○	⊙
Hygienicity	×	△	□	□	△	△	○	⊙	⊙
Heat resistance	×	×	△	△	△	○	□	○	⊙
Flavor durability in processed food	×	△	△	△	△	△	○	○	⊙
Color tone in food	×	×	○	○	○	○	○	○	○
Stability of quality and price	×	×	○	○	○	○	○	○	○

⊙: Superior; ○: slightly superior; □: intermediate; △: slightly inferior; ×: inferior.

three types: water-soluble single layer coating, water-insoluble single layer coating, and double layer. The first is produced by emulsifying an oleoresin or essential oil in a water solution of high molecular substance, such as dextrin, sodium caseinate, and gum arabic, followed by spray-drying into powder. The second is produced by mixing oleoresin or essential oil in heat-melted solid fat, followed by spraying and cooling. The third is produced by a more complicated procedure. Oleoresin or essential oil is well mixed in heat-melted solid fat, and this mixture is emulsified in a previously emulsified solution containing essential oil or oleoresin, followed by spray-drying into powder form. The major characteristic of this coating spice is that no microorganisms can survive in it and it can be preserved for long periods of time. Double layer coating spices, which are covered with both minute solid fat particles and water-soluble high molecular substances, can be preserved for especially long periods of time.

Many spices are milled to small particles before they are used for cooking. Particle size affects both flavor and pungency of a spice. Mori studied this effect using cardamom and white pepper [2]. He milled both spices to powders with different particle sizes powder to be used in sausage, then checked how spice flavor differed according to particle size. As shown in Table 3.3, flavor was significantly stronger when the medium ground spice (28 mesh) was used than when the finer ground spice (80 mesh) was used. The coarse ground spice in this study (14 mesh) seemed to be too large to be uniformly distributed so that no significant difference could be found, although there was a tendency for more flavor when coarse ground spice was used. Mori also examined emulsified spices and coating spices to compare their deodorizing effect on fish smell. He used granules coated with hardened oil as the coating spice and an emulsion made with gum arabic as the emulsified spice. Each was used in sausage to evaluate organoleptically how effectively the fish smell was neutralized. The emulsified spice was so dispersed that it suppressed the fish smell much more than the coating spice. But the flavor and pungency of the coating spice was rated better than those of the emulsified spice.

TABLE 3.3 Difference in Flavor and Pungency Due to Granulation of Black Pepper

Ratio	White pepper (0.2%)	Cardamom (0.1%)
A:B	B = 23/30	B = 21/30
B:C	B = 16/30	B = 18/30
A:C	C = 18/30	C = 17/30

A: Fine-ground, 80 mesh; B: ground, 28 mesh; C: coarse-ground, 14 mesh.
Source: Ref. 2.

Supercritical fluid extraction (SCFE) has recently been applied to flavor extraction from spice. The major characteristics of this method is the use of carbon dioxide gas as an extract solvent in place of the organic solvent used in case of solvent extraction. Steam distillation and organic solvent extraction have both been widely used to extract compounds from spices. However, these two extraction methods both tend to degrade the original spice flavor and lose volatile compounds due to the high temperatures used. On the other hand, since extraction temperatures in the SCFE method can be low, a high-quality extract with good flavor can be obtained. The characteristics of supercritical carbon dioxide used as an extract solvent are as follows:

1. Its density is close to that of a liquid, but its viscosity is very near to that of a gas and its diffusion ability is 100 times that of a gas.
2. Both the critical temperature and the critical pressure are relatively low.
3. The simple operation for temperature and pressure enables one to easily control the fluid density, making this extraction method superior in selectivity, such as separation.
4. Since this fluid is inactive, in addition to the fact that extraction operation can be conducted at relatively low temperature, the extract is not greatly decomposed or oxidized during the extraction process.
5. The ignitability of this fluid is low, and it is nonpoisonous, odorless, and tasteless.
6. This fluid does not remain in the extract.
7. Carbon dioxide is inexpensive and easy to obtain.

Figure 3.2 shows the difference in extracted spice compounds according to extraction method: steam distillation, ethanol extraction, methylene chloride, and SC carbon dioxide gas [3]. This figure illustrates that steam distillation can separate essential oils, but not higher terpenes, free fatty acids, fats, wax resins, or colorants. It also shows that alcohol can extract parts of some higher terpene compounds, although some flavor top notes disappear during the evaporating process. On the other hand, many compounds can be extracted with both methylene chloride and supercritical carbon dioxide. Methylene chloride extraction has the disadvantage of scattering some flavor compounds during the evaporating process, whereas supercritical carbon dioxide can extract essential oils at relatively low temperatures, and higher terpenes, free fatty acids, fats, waxes, and colorants could be extracted accordingly as the extract temperature is increased. When this technique is applied to the extraction of spice, there are a couple of advantages. The typical spice flavor, which is usually very sensitive to high temperature, can be extracted without much damage because the temperature used for this extract method is so low. This method also can fractionate "specific components" such as color and specific flavors by controlling extract temperature and pressure. For example, when a color component is extracted from a spice, with

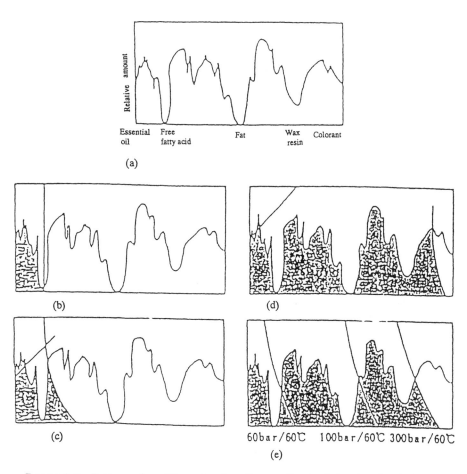

FIGURE 3.2 Patterns for different extraction methods: (a) natural substance; (b) steam distillation; (c) ethanol; (d) methylene chloride; (e) carbon dioxide. Shaded area indicates compounds extracted. (From Ref. 3.)

solvent extraction some flavor components and others besides the expected color itself may be obtained, while the extract obtained with the SCFE method will contain only the target color component.

Although there are still a couple of disadvantages of SCFE for spices, such as a high processing cost, it has many unique characteristics. The use of this method is expected to increase as more research into "specific" components of each spice is conducted, including physiologically active compounds.

III. THE EFFECT OF TIMING ON FLAVOR

Most of the flavor compounds in spices are contained in their volatile oils and are changed or damaged upon heating. The timing of adding a spice when cooking is, therefore, considered one of the most important aspects of making full use of spices during cooking. The cooking process can be divided into three parts: (1) preparation before heating, (2) cooking using heat, and (3) final preparation (after the food is removed from heat).

As an example of the importance of the time chosen to add a spice during cooking, we will use the example of black pepper. This spice can be sprinkled over meat with salt before baking or barbecuing. It can be added to a sauce or stew during the "actual" cooking period, and it can be sprinkled over beef after it has been removed from the heat. Table 3.4 shows how the flavor of food can be changed depending upon the timing for adding spice.

Most components of essential oils in spices are sensitive to heat. A spice will lose less of its delicate aroma if it is used after the food has been removed from the heat. The timing for adding a spice is especially important when it is used for the purpose of deodorizing fish or meat. Because the deodorizing effect of a spice against a fish or meat smell is due mainly to some chemical reaction, it is important to determine the appropriate time during the cooking process to add the spice. Since, in general, it is necessary to use a relatively high temperature for spice components to react with these smelled compounds, spices used for their deodorizing properties should be added either during the heating process or even before heat is applied.

For example, garlic is often used for the purpose of deodorizing the smell of meat. If grated garlic were sprinkled over meat already baked or barbecued, one would experience a strong garlic smell. However, if used either during heating or before the heating process, garlic works effectively to get rid of the meat smell. In this case, the heating process has the effect of reducing the typical strong garlic smell as well as deodorizing the meat smell, and a relatively large amount of garlic can be used. Beef is sometimes baked with garlic slices in a steak restaurant because this process neutralizes the meat smell and changes the garlic flavor to a milder, roasted one.

TABLE 3.4 Flavor Change According to Time Spice Added

Spice	Time spice added		
	Before cooking	During cooking	After cooking
Black pepper	Weak flavor	Slightly weak flavor	Strong flavor
Clove		A little weak flavor	Strong flavor
Onion		Weak or no flavor	Strong flavor

IV. THE EFFECT OF BLENDING ON SPICE FLAVOR AND AROMA

Statistically, there are very few cases where only one kind of spice is used for cooking. In most cases, a spice is used in combination with other one or more kinds of spice. For example, when bouillon or soup is prepared by extraction from beef, pork, or chicken bones, many spices are used for the purpose of deodorizing their typical smells. In general, four or more spices such as bay leaves, thyme, celery, and parsley are added together during extracting process for these meats. This "grouping" technique of spice application is also used in processed foods, such as ketchup or Worcestershire sauce.

It is difficult to determine the amount of a certain spice to use for a cooking because of the differences in personal preferences. Even a pleasant flavored spice can have a medicinelike smell or taste if used to excess. But such a medicinelike flavor or aroma of one spice can be reduced by working by blending with other spices. Each spice has a typical flavor, the quality of which cannot be changed when only that spice is used for cooking, although the "strength" of its flavor can be controlled by adjusting the amount to be added. However, if a spice is grouped with other spices, the total spice combination will have a more delicate flavor than when each is used individually. This is called the "spice-blending" effect.

When this blending technique is used for the purpose of reducing the specific flavor of one spice, it is most effective to use a spice having similar flavor characteristics. For example, the addition of celery or celery salt to tomato juice can eliminate the typical "greenish" flavor of tomato because celery has a similar flavor. Besides celery, carrot and parsley are often added to tomato juice or tomato-based vegetable juice to make it more drinkable. One of the advantages of this spice-blending technique is that the amount of spice to be added can be increased to decrease the "undesirable" typical flavor. Most people who dislike spices, especially children, say that they do not like strong spice flavors. Spice-blending techniques can solve this problem.

The blending effect can be easily confirmed by comparing alcoholic beverages containing different combinations of spices. The typical spice flavor of an alcoholic drink containing only one kind of spice tends to be perceived as stronger than that of a drink to which three or more spices have been added. In this case, whole spices tend to show the blending effect more clearly than ground spices. For example, if clove flavor in a brandy containing only clove is compared with that in another brandy containing not just clove but also cinnamon and allspice, the former would have a stronger clove sensation than the latter because of the similar flavors of cinnamon and allspice and clove—both cinnamon and allspice would mask the typical flavor of clove.

The many kinds of "blended" spices available worldwide (Table 3.5) include chili powder, used in Mexican cooking, quatre epices and bouquet garni, used in French cooking, and herb seasoning, used in Mediterranean cooking.

Japanese shichimi (seven spice) and Chinese five spice powder are also blended spices.

The spice-blending technique involves combining spices from the same botanical family. This technique results in a well-rounded flavor. For example, Perilla family herbs are often blended together to produce Italian herb mix or pizza spice mix.

V. THE EFFECT OF AGING ON SPICES

The flavor of blended spices becomes milder, or "rubbed," after they are stored for a period of time. Immediately after blending, the flavor of the spice mixture is a kind of "assembly" of separate flavors. However, as time passes, the separate flavors incorporate into a single flavor. This is called the aging effect.

Aging techniques are widely used to produce whiskey, miso, and other products (especially fermented foods), and a special aging method is devised for each product. This aging process is very important, especially in creating the special flavors of wine and whiskey. These alcoholic beverages are aged in wooden barrels under controlled temperature for years. A wine aged for a longer period of time tends to be of higher quality.

Curry powder and chili powder, the most well-known blended spices, acquire their characteristic flavor qualities with the aging process. For most manufactured curry or chili powders, aging conditions such as temperature and aging period are not specifically determined. The aging effect would be obtained simply by leaving the blended spice in its container for several months.

The aging effect is thought to be due to both gradual chemical and physical changes of the essential oils of each spice. Therefore, it is better to age a blended spice in a cool, dark place than at warmer temperatures because the volatile flavor components would change slowly under mild conditions. Ultraviolet light and high temperatures, which deteriorate the properties of the essential oils of spice, should be avoided. Curry powder, for example, is better roasted in a pan at low heat if its flavor is kind of "rough." This process actually promotes the aging effect, whereas roasting at too high a temperature will evaporate the volatile compounds making up the typical curry flavor and result in a product with poor flavor. The roasting process is also utilized in most spice companies for manufacturing curry powder, not only to promote the aging effect but also to create a delicate flavor characteristic specific to each curry product.

VI. SYNERGISTIC AND SUPPRESSIVE EFFECTS OF SPICES

The taste of a food can often change when it is combined with other food or beverages. A synergistic taste effect is when the taste of one feed component is enhanced by association with other food components. A suppressive effect is

TABLE 3.5 Various Kinds of Blended Spices Found Worldwide

Spice	Curry powder	Garam masala	Chili powder	Bouquet garni	Pickling spice	Quatre épices	Italian herb	Chinese five spice	Japanese seven spice
Aromatic									
Garlic	○	○	○						
Onion	○		○						
Bay leaves	○			○			○		
Clove	○	○	○	○	○	○		○	
Nutmeg	○	○		○	○	○			
Perilla									○
Thyme		○	○	○		○			
Rosemary						○			
Caraway	○	○	○	○					
Sage							○		
Oregano			○	○			○		
Savory		○		○			○		
Coriander	○			○	○				
Parsley	○			○					
Celery	○	○							
Cinnamon						○		○	
Allspice			○		○				
Dill			○		○				

Cumin	○	○	○		○				
Marjoram				○			○		
Star anise								○	
Basil				○			○		
Orange peel								○	○
Mace	○				○				
Fennel	○	○		○	○			○	
Sesame	○								○
Poppy seed	○								○
Hemp									○
Fenugreek	○	○			○				
Cardamom	○	○			○				
Pungent									
Pepper	○	○			○	○		○	
Red pepper	○	○	○		○	○			○
Mustard	○				○				
Ginger	○	○			○	○			○
Japanese pepper								○	○
Colorant									
Paprika			○						
Turmeric	○								
Saffron	○			○					

when a certain taste is decreased in its strength by combining with other components (also known as the "offset effect").

It is well known that sodium glutamate exhibits a remarkable synergistic effect with inosinic acid. This phenomenon is known to be a kind of taste illusion. Similar phenomena can be observed in the case of some spices. One example is the synergistic effect of ginger used in soft drinks. Ginger has a typical earthy smell, but it has fresh citrus flavor and a fresh stimulus as well. These enhance the "freshness" of some soft drinks when ginger is added to them. Cinnamon has the effect of enhancing the sweetness of sugar-containing foods and has been widely used in cooking baked goods and confections. Cinnamon contains a sweet aroma but does not contain any substance expressing a sweet taste. But when it is combined with a sweet food, the sweet sensation is enhanced because of the synergistic effect between the sweet taste of sugar and the sweet aroma of cinnamon. On the other hand, when the salt concentration in a soup exceeds a certain level (usually around 1%), addition of black or white pepper is thought to suppress the salty taste. This phenomenon will be explained further later in this chapter.

The above are examples of synergistic and suppressive effects occurring between spices and foods. Synergistic effects have also been observed among different kinds of spice. Take, for example, the addition of vanilla to fruit cake. Vanilla essence is made by combining alcohol with the essential oil of the vanilla bean to make it soluble in water. But, the flavor of vanilla, when used for baking, is weakened when heated to over 100°C, causing volatilization of its essential oil. In this case, adding a small amount of clove or nutmeg will create the impression of a stronger vanilla flavor because the faded flavors of these spices when heated are close to that of vanilla. Synergistic and suppressive effects of some spices are listed in Table 3.6.

Ito et al. [4], reporting on how spice suppresses mutton odor, made clear how the synergistic or suppressive effects of spices occur. They added a phased concentration of diluted solution of an individual spice to an emulsion of 1% mutton oil and determined organoleptically the amount of diluted solution of spice necessary to deodorize the smell of mutton. They also used dilute mixed solutions

TABLE 3.6 Synergistic and Suppressive Effects of Spice

Food	Combination	Change of flavor perceived	Effect
Ice cream	Sugar + vanilla	Sweetness: stronger	Synergistic
Ice cream	Sugar + pepper	Sweetness: weaker	Suppressive
Cake	Sugar + cinnamon	Sweetness: stronger	Synergistic
Salad dressing	Salt + pepper	Saltiness: stronger	Synergistic
Soup	Salt + pepper	Saltiness: weaker	Suppressive
Fruitcake	Vanilla + clove	Vanilla flavor: stronger	Synergistic

Cooking with Spices

TABLE 3.7 Deodorizing Effect of Various Spices

Order of deodorizing effect	Spice	Deodorizing point (×100)
1	Sage	0.7
2	Thyme	2.5
3	Clove	3.0
4	Caraway	4.0
5	Coriander	5.0
6	Garlic	23.0
7	Celery	25.0
8	Cardamom	30.0
9	Allspice	30.0
10	Nutmeg	45.0
11	Cinnamon	50.0
12	Ginger	90.0
13	Onion	190.0
14	Pepper	600.0

Source: Ref. 4.

of two or more kinds of spice to observe their deodorizing effects. One spice with a strong deodorizing effect is sage, followed by thyme, clove, and caraway (Table 3.7). Table 3.8 shows the relationship between the calculated amount of two or more spices necessary to deodorize mutton smell and the amount actually added. In the case of a mixture of sage and thyme, which showed the strongest and second strongest effect, respectively, the actual amount necessary is more than the calculated amount. It is the suppressive effect of both spices. They categorized the effects of two or more spices into three effects: synergistic, suppressive, and arithmetic (Table 3.9).

VII. COOKING WITH SPICES TO REDUCE SALT AND SUGAR CONTENT

Judicious use of spices often enables one to reduce the amount of sugar or salt used for cooking. The more healthful foods produced could be taken advantage of in settings such as hospitals.

Various characteristics of food are perceived through the nose, mouth, eye, and fingers. The sense of taste is a general term of the perception experienced when a food is observed and taken into the mouth and then into the stomach. This sense has three aspects—psychological, physical, and chemical—and spices can influence these aspects.

TABLE 3.8 Synergistic Deodorizing Effect of Spice Combinations

Order of deodorizing effect	Spice combination	Deodorizing point		Calculated value minus actual measurement
		Calculated value	Actual measurement	
1	Sage + Clove	1.85	1.40	0.45
2	Sage + Caraway	2.85	2.20	0.65
3	Sage + Coriander	2.85	2.85	0.00
4	Sage + Celery	12.85	12.70	0.15
5	Sage + Garlic	16.85	80.00	63.15
6	Sage + Thyme	1.60	170.00	168.40
7	Caraway + Celery	14.50	12.07	2.43
8	Thyme + Clove	2.75	2.75	0.00
9	Coriander + Garlic	14.00	14.00	0.00
10	Clove + Celery	14.50	14.50	0.00

Source: Ref. 4

Spices are natural substances containing sweet compounds, bitter compounds, and others. The major characteristic of a spice is its "pungent" sensation. In addition to several pungent compounds, many aromatic compounds at high concentration produce a physical stimulus, which can be perceived as a pungent sensation.

Sweet, tart, salty, and bitter tastes are generated by substances that dissolve in water and stimulate certain nerves (Table 3.10). The components of a spice are perceived as a physical stimulus in the mouth. The palatability of food is influenced not only by its taste, smell, and texture, but also by psychological factors perceived through visual sensation, like its color and shape. Of the various factors that determine the palatability of food, those that spices can influence directly include pungency, flavor, and color.

One's sense of taste changes with age. Table 3.11 shows the minimum concentration of each basic taste perceived within each age category [5]. As indicated in the table, higher minimum amounts for sugar, salt, and sulfate quinine were required for detection by the 75- to 89-year age group than by the 15- to 29-year age group. This indicates that taste sensitivity worsens with age, which can be explained by a lower number of a taste buds, which are the taste sensors. It is said that the number of taste buds existing in one vallate papillae drops suddenly at about the age of 70. Using spices to enhance the flavor of foods while reducing the salt or sugar content could be especially useful in light of certain diseases of adults, including diabetes or kidney disease.

The sensation of saltiness is created by a chemical reaction occurring on the tongue, but the pungent sensation caused by some spices is a physical sense

TABLE 3.9 Deodorizing/Masking Effect of Various Spices

	Sage	Thyme	Clove	Caraway	Coriander	Garlic	Celery
Sage		× −168.40	○ 0.45	○ 0.65	△ 0	× −63.15	○ 0.15
Thyme	× −168.40		△ 0	○ 0.01	○ 0.59	○ 3.95	○ 5.74
Clove	○ 0.45	△ 0		× −176.50	○ 0.25	○ 0.05	△ 0
Caraway	○ 0.65	○ 0.05	× −176.50		○ 0.25	△ 0	○ 2.43
Coriander	△ 0	○ 0.59	○ 0.25	○ 0.25		△ 0	○ 5.50
Garlic	× −63.15	○ 3.75	○ 0.50	△ 0	△ 0		× −166.50
Celery	○ 0.15	○ 5.74	○ 0	○ 2.43	○ 5.50	× −166.00	

○: Synergistic; △: suppressive; ×: arithmetic.
Source: Ref. 4.

TABLE 3.10 Bitterness Compounds in Spices

Spice	Bitterness compounds
Allspice	Eugenol
Bay leaves	Linalool, polyphenol compounds
Caraway	Carvone
Cardamom	Terpineol
Cinnamon	Eugenol, citral
Clove	Eugenol, polyphenol compounds
Coriander	Linalool
Cumin	Cumin aldehyde
Dill	Carvone
Marjoram	Terpineol
Nutmeg/mace	Eugenol
Oregano	Carvacrol, thymol
Rosemary	Borneol
Sage	Linalool
Thyme	Thymol

thought to be similar to pain. It is, therefore, generally considered to be difficult to replace salt completely with the pungent sensation of spice. But one can expect the amount of salt necessary to be reduced if the pungency, flavor, and coloring effects of spices are used well. For example, using a pungent spice or spice leaves in a salad can make it very tasty. Also, using colored spices may enhance the total deliciousness of a salad. Such dishes can be called low in salt because the total amount of salt used is decreased by utilizing spices.

Ohta investigated the influence of salt on salty sensation [6]. He prepared five salt solutions with concentrations of 1.38, 1.20, 1.04, 0.91, and 0.08%,

TABLE 3.11 Minimum Concentration of Solutions Tasted by Age Group

Solution	15–29 (N = 25)	30–44 (N = 16)	45–59 (N = 23)	60–74 (N = 27)	75–89 (N = 9)
Sucrose	0.54 (g/100 ml)	0.522	0.604	0.979	0.919
Salt	0.071	0.091	0.11	0.027	0.31
Hydrochloric acid	0.0022	0.0017	0.0021	0.003	0.0024
Sulfate quinine	0.000321	0.000267	0.000389	0.000872	0.00093

Source: Ref. 5.

respectively, and also prepared two salt-pepper solutions by adding white pepper at 0.08 and 0.16% to the 1.04% salt solution, comparing the salty sensation of each salt solution with that of each salt-pepper solution organoleptically using several panels. Table 3.12 shows the results of test panels asked to determine which salt solution was equivalent in salty sensation to the 1.04% salt/0.08% pepper solution (solution A). In the first attempt, 6 of 11 panels said that the salty sensation of this salt-pepper solution was equivalent to the 1.04% salt solution. One panel each said that its saltiness was equal to that of the 1.20 and 0.91% salt solutions, and 3 determined the 0.80% salt solution to be equivalent. Table 3.12 shows the test results of three attempts, including the first attempt. The assumed salt concentration of solution A containing salt and pepper was calculated to be 0.97% at the first attempt. This numerical figure was calculated using the following information and equation:

Actual salt concentration (%)	Concentration point
0.80	−2
0.91	−1
1.04	0
1.92	+1
1.93	+2

$$1.04 - (1.04 - 0.91) \times \frac{\Sigma \,(\text{concentration point})(\text{panel number})}{\text{total panel number}}$$

As shown in the above table, the average value of the assumed salt concentration of solution A was 1.01%. But when white and black pepper were added to the 0.8% salt solution, the assumed salt concentration increased to 0.96% and 0.91%, respectively. These test results show that salty sensation is enhanced by

TABLE 3.12 Effect of White Pepper on Salty Taste[a]

| Salt | 1.38 | 1.20 | 1.04 | 0.91 | 0.08 | | | Equivalent |
Salty taste	2	1	0	−1	−2	Total	Average	salty taste (%)
First attempt	0	1	6	1	3	−6	−0.54	0.97
Second attempt	0	1	10	0	0	1	0.09	1.05
Third attempt	0	0	9	2	0	−2	−0.18	1.01
							Average	1.01

[a]Eleven panels were asked individually which salt solution was equivalent to salt-pepper (0.08% white pepper and 1.04% salt) solution.
Source: Ref. 6.

TABLE 3.13 Effect of Pepper on Salty Taste

Salt (%)	White pepper (0.16%)	Black pepper (0.16%)
0.80	0.91	0.96
0.91	1.02	0.95
1.04	1.10	1.09
1.20	1.20	1.15
1.38	1.35	1.28

Source: Ref. 6.

adding pepper when the salt concentration is low, but there is almost no effect if salt concentration exceeds 1.0 percent (Table 3.13).

Goto et al. [7] studied the effect of capsaicin on the preference for salty taste. They first observed how intake of capsaicin influenced the concentration of salt preferred by test rats, and then checked the chorda tympani to see how capsaicin influenced the reception of salt.

In the first experiment, rats used for the test were divided into two groups. The rats in one group were fed a capsaicin-containing diet, the others a noncapsaicin diet. Saltwater with 0.5, 0.9, and 1.4% concentrations were prepared. One-tenth of 8-week-old rats fed with noncapsaicin diet were observed to drink the 1.4% saltwater, while the rats fed the capsaicin-containing diet did not drink this saltwater during the 8-week testing period (Fig. 3.3). The average amount of salt

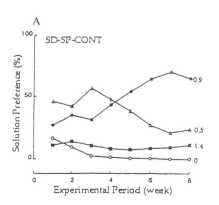
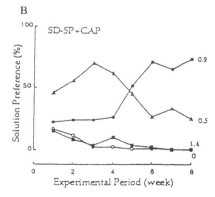

FIGURE 3.3 Effect of capsaicin on preference of rats for noncapsaicin or capsaicin diet with four different salt concentrations. SD male rats were given (A) noncapsaicin diet or (B) capsaicin diet (3 mg/rat/day) for an 8-week test period. The numerical values indicate salt concentration in each diet. (From Ref. 7.)

consumed by rats fed the noncapsaicin diet for 8 weeks was 7 g/100 g body weight, while the amount consumed by capsaicin fed rats was 5 g/100 g body weight. Goto et al. tried to confirm this fact by feeding capsaicin directly into the stomach (3 mg/rat/day) for 10 days, since they thought that the change in preference with capsaicin might be due to paralysis of the tongue. But it was found that rats ingesting capsaicin tended not to prefer high-concentration saltwater. Investigation of the chorda tympani showed a result similar to that obtained with the above experiments. The chorda tympani of the rats ingesting capsaicin reacted to 0.5% saltwater, while rats ingesting normal diets reacted to 1.0% saltwater. It means chorda tympani of rats have been stimulated twice by ingested capsaicin. Other research studied SHR-SP rats with high blood pressure, preferring twice as much salt as normal rats, and confirmed that the amount of salt taken was decreased when capsaicin was fed, resulting in less of an increase in blood pressure. These results suggest that capsaicin can reduce the preference for salt and, therefore, the amount of salt used in cooking.

In the same manner, the sweet aroma of some spices can increase the sweet sensation, allowing one to decrease the amount of sugar used in some kinds of foods. Many chemical compounds have a sweet taste. Sucrose and glucose, important energy sources, are the most popular sweet compounds. The sweet sensation is affected not only by the amount of sugar but also by the texture of the food. Sweetness needs to be stronger for foods that are hard in texture, such as caramel or other sweet confections, than for soft-textured foods such as soft drinks or ice cream.

Sugars are classified roughly into mono-, di-, tri-, oligo-, and polysaccharides. The sweetness of each of these sugars varies, although their chemical structures are similar. All monosaccharides have a sweet taste. Most disaccharides, such as sucrose, have a sweet taste as well, but some disaccharides have just a slight sweetness. Polysaccharides have almost no sweet taste. In addition to many types of sugar, sugar alcohol, amino acids, peptides, and others give a sweet sensation, but the characteristics of these compounds, including sugar, differ. For example, some have a slight acid taste, while others are slightly bitter. There is also a difference in the time it takes to perceive sweetness: sometimes it is perceived immediately upon tasting, sometimes as more of an aftertaste. Temperature is also a factor in the perception of the sweet sensation. Figure 3.4 shows the change in sweetness of different sugars at different temperatures. In this figure it is assumed that the sweetness of sucrose at each temperature is 100, and the sweetness of each type of sugar is compared with that of sucrose. You can see from this figure that fructose is 1.4 times as sweet at 0°C but 0.8 times as sweet at 60°C as sucrose at the same temperatures. It is difficult to increase the sweet sensation of ice cream because the sweetness of sucrose is weak at low temperatures and adding extra sucrose changes the balance of total taste and flavor. In this case, it is more effective to add vanilla essence, which has no sweet taste but does have a

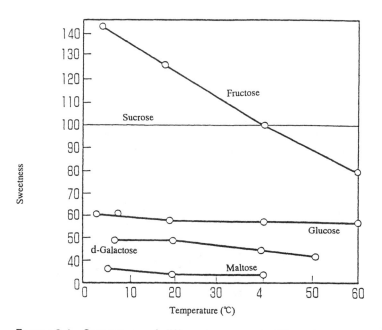

FIGURE 3.4 Sweetness of different sugars at different temperatures. (From Ref. 6.)

sweet aroma, than to add extra sucrose because an aroma of vanilla can enhance the total sweet sensation spread into the mouth synergistically with the taste of sugar compounds. Spices in general contain several sugar compounds, and some spices like onion and garlic actually have a sweet taste (Table 3.14). Table 3.15 lists the chemical compounds that possess a sweet aroma contained in spices. Anetol, one of the compounds having a strong sweet aroma, is contained in anise seed, fennel seed, and star anise, all of which can be used for enhancing the sweet sensation of foods. Cinnamon and cardamom are also utilized in sweet foods.

TABLE 3.14 Sugars Contained in Some Spices

Spice	Sugar
Onion, garlic	Glucose, fructose, sucrose, lactose, raffinose, scorodose
Clove	Glucose, rhamnose, xylose, galactose, fructose, lactose
Oregano, sage	Mannose, glucose, fructose, rhamnose, xylose, arabinose, galactose, lactose, maltose, sucrose, raffinose

TABLE 3.15 Sweet Flavor Compounds in Some Spices

Spice	Sweet flavor compounds
Anise	Anethole
Fennel	Anethole
Star anise	Anethole
Basil	Methyl chavicol, anethole
Cinnamon	Cinnamaldehyde
Vanilla	Vanillin

VIII. THE DEODORIZING/MASKING EFFECTS OF SPICES

Spices are sometimes used for the purpose of deodorizing or masking the smell of raw materials. The flavor of a spice does not itself come into play in this case as much as the ability of the spice to mask an undesirable odor.

The desirability of such a masking effect often depends upon the population in question. For example, Japanese people tend to dislike the typical smell of soymilk, and soymilk products sold in Japan are usually deodorized by some means. Chinese people, on the other hand, do not like soymilk products from which the smell has been completely removed.

Deodorizing/masking can be divided into three types: chemical, physical, and sensation (Fig. 3.5). In chemical deodorizing, smelled compounds are changed to nonvolatile compounds or to odorless substances through some chemical reaction such as neutralization, oxidization, or reduction. A fish smell can be deodorized by soaking it in lemon juice or vinegar. The deodorizing mechanism for this phenomenon involves alkali trimethylamine, one of the odorous compounds, being neutralized by the acid to become a nonvolatile compound. In physical deodorizing, unpleasant smelling compounds are absorbed by porous active carbon or zeolite. Active carbon is also utilized as a refrigerator deodorant because of this absorbing effect. Sensational deodorizing is theoretically divided into two types: "masking" in a narrow sense, in which a strong spice flavor covers an unpleasant smell, and "offset deodorizing," in which two compounds having different odors become odorless when mixed. In this technique, the actual odor does not decrease in intensity or disappear—it is simply not perceived as readily.

In addition to these techniques, some spices have a fourth deodorizing activity: the mucous membrane inside nose, which works as a sensor for smell, is actually paralyzed so that the odor is not perceived. Spices having this kind of deodorizing activity include those having volatile pungent compounds, such as wasabi or mustard. Mustard is often used in fermented soybeans called "Natto," a

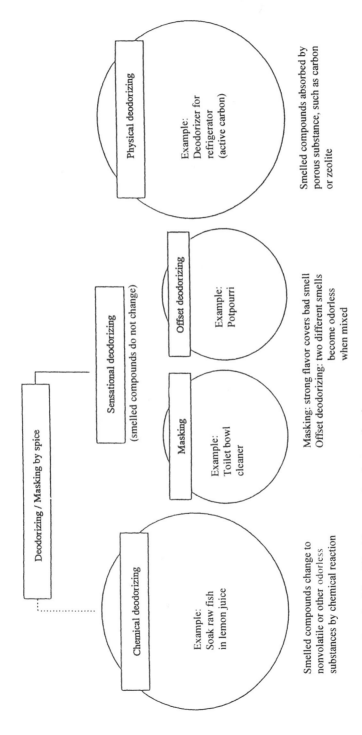

FIGURE 3.5 Deodorizing/Masking mechanisms of spices.

very popular side dish in Japan, and wasabi for raw fish. Their deodorizing effect is achieved via isothiocyanate compounds, which are flavor and pungent components of wasabi and mustard, which stimulate the mucous membrane of the nose when ingested. The secreted mucus paralyzes the sensor function, resulting in a temporary inability to perceive any odors.

We would also like to mention some practical techniques to enhance the effectiveness of spice use to deodorize unpleasant odors. Bacteria existing in the mouth decompose saliva and food sediment into bad smelling compounds, including volatile sulfur compounds, volatile nitrogen compounds, low fatty acids, alcohol, aldehydes, and acetone, all of which cause bad breath. The actual amounts of volatile sulfur compounds, such as hydrogen sulfide, methyl mercaptan, and dimethylsulfide, generated are proportional to the strength of the bad breath. Recent research revealed that the concentration of methyl mercaptan correlates more closely with the bad breath than the other compounds, showing that the major cause of bad breath is methyl mercaptan. Several means have been considered for deodorizing bad breath: (1) pasteurizing bacteria in the mouth, (2) seizing methyl mercaptan with a deodorizer (chemical deodorizing, physical deodorizing), and (3) masking the bad smell with some flavor compounds. We will discuss the deodorizing means of removing interoral odor below.

Sodium copper chlorophyllin (chlorophyll) is known to have a chemical deodorizing function and is often used in breath-freshening chewing gum and some kinds of oral-care products such as toothpaste. Tokita et al. [8] studied the deodorizing effects of various plants, including spices, against methyl mercaptan. They tried to find a plant with a stronger deodorizing effect than sodium copper chlorophyllin [8]. Methyl mercaptan is considered to be major compound of bad breath and is one of the factors to be considered for developing canned foods because it is sometimes generated when heated strongly. Table 3.16 shows how much methyl mercaptan would be captured by the methanol extract of certain spices. Only a few kinds of spice showed a deodorizing effect greater than sodium copper chlorophyllin. Thyme has a deodorizing effect almost as strong as sodium copper chlorophyllin, while the effect of rosemary was found to be six times and that of sage three times stronger than that of sodium copper chlorophyllin (Fig. 3.6). The effective deodorizing compounds of rosemary and sage are carsonol and rosemanol that also have an antioxidant function. These compounds when isolated were confirmed to be many times stronger in deodorizing function than sodium copper chlorophyllin. These authors also found that the major deodorizing compound was the decolorized and de–essential-oiled ethanol extract section. That is to say, the nonvolatile section was also found to seize methyl mercaptan.

According to the Weber-Fecher law, the strength of an odor perceived by the sense of smell is proportional to the logarithm of the concentration of the smelled compounds. In other words, the sensational strength perceived with the five senses is proportional to the logarithm of the actual strength of these stimuli. For example, even if 99% of the total smelled compounds is eliminated chemically, the sensational strength perceived is reduced only 66%. It is, therefore, more

TABLE 3.16 Deodorizing Rate[a] of Spice Extracts

Family	Spice	Deodorizing rate (%)
Labiatae	Sage	95
	Thyme	99
	Savory	90
	Oregano	93
	Marjoram	91
	Rosemary	97
	Basil	63
	Perilla	91
	Mint	90
Umbelliferae	Celery	44
	Dill	13
	Anise	27
	Cumin	11
	Fennel	0
	Caraway	24
	Coriander	3
Myrtaceae	Clove	79
	Allspice	61
Rutaceae	Japanese pepper	72
Compositae	Tarragon	36
	Chrysanthemum	12
Solanaceae	Paprika	8
Leguminosae	Fenugreek	4
Piperaceae	Pepper	30
Zingiberaceae	Turmeric	5
	Ginger	4
	Cardamom	9
Magnoliaceae	Star anise	39

[a]Deodorizing rate—percent of methyl mercaptan (500 ng) captured by methanol extract of each spice.
Source: Ref. 8.

effective to use an aromatic spice to deodorize the remaining 1% via the masking function.

Judging from the facts described above, the spices in the Labiatae family, including rosemary, are most effective at deodorizing because they have both a chemical and a sensational deodorizing function and enhance the deodorizing/masking effect very effectively. Several reports have tested the deodorizing effectiveness of various spices against a variety of bad smelling compounds. It is now common knowledge that trimethylamine oxide is reduced to trimethylamine due

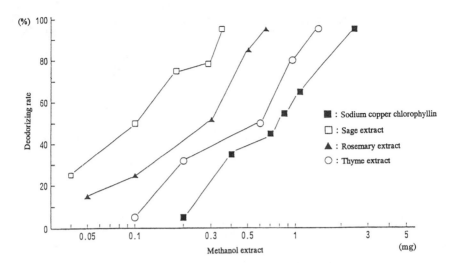

FIGURE 3.6 The deodorizing effects of different spice extracts. (From Ref. 8.)

to microbial action, and much research on this in connection with the freshness of seafood has been reported [9–12]. Shimizu et al. [12] used trimethylamine as a reverse indicator of seafood freshness because it increases as freshness is lost [12]. In this study, it was confirmed that trimethylamine and ammonia increased as spoilage advanced in fish, while trimethylamine oxide gradually decreased.

Kikuchi et al. [13] evaluated the masking effect of each spice sensorially by adding a spice solution with phased concentration to a trimethylamine solution. According to their report, sensory evaluation found the essential oils of onion, bay leaf, and sage to be the most effective in suppressing the odor of trimethylamine, followed by those of thyme, caraway, ginger, clove, and cassia. It was also confirmed that the amount of trimethylamine quantitatively analyzed by gas chromatography was reduced when the essential oils of sage and other effective spices were mixed with trimethylamine (Table 3.17). Table 3.18 shows that trimethylamine decreases gradually as a function of time elapsed when the essential oils of ginger and sage are added. Based on the research of Kikuchi et al., Niwa et al. [14] chose the essential oils of onion and bay leaf, which were also expected to suppress trimethylamine, to conduct further research on their deodorizing/masking effect. They found that the phenol partition of the essential oil of bay leaf was very effective in masking trimethylamine. Further refining of this phenol section revealed that the effective compounds were eugenol and p-allyl-phenol. Onion, however, which was confirmed sensually to suppress trimethylamine, did not show any change on gas chromatography. Shiraishi et al. [15] conducted research on the suppressive compounds in some fermented seasonings against fish

TABLE 3.17 Amounts of Trimethylamine (TMA) Detected When Mixed with Certain Essential Oils

Spice	TMA (mm^2)	%	Spice	TMA (mm^2)	%
Control	333	100	Clove	288	86.4
Bay leaves	312	93.7	Thyme	288	86.4
Ginger	300	90.1	Caraway	276	82.9
Nutmeg	297	89.2	Pepper	249	74.8
Cinnamon	294	88.2	Sage	222	66.7

Source: Ref. 13.

smell, and presumed that this suppression was due to the change of flavor compounds to nonvolatile ones caused by formation of a π-complex as well as salt. They examined the suppressive effect of phenol compounds, including vanillin and terpene compounds, against trimethylamine and found that some compounds such as vanillin register agreement between the results of sensory evaluation and the decrease in the amount of trimethylamine, whereas other compounds such as acetovanillone do not (Table 3.19). Some other compounds such as anisole were found to be effective organoleptically in spite of no decrease in the amount of trimethylamine measured, indicating the existence of a suppressive mechanism against bad smell besides that of salt formation (Table 3.20) [16].

The deodorizing/masking effect of vanillin is assumed to be due to its bonding with fish smell compounds, since a vanillin solution absorbs trimethylamine very well. Kawamura et al. evaluated cooking aromas from heated chicken

TABLE 3.18 Change in Amount of TMA in TMA-Spice Mixture

	Peak area of TMA (mm^2)	
Day	Ginger added	Sage added
0	700	712
1	650	540
2	600	450
3	550	—
4	495	225
6	—	75
7	440	45

Source: Ref. 13.

TABLE 3.19 Odor-Masking Effect on TMA
of Phenolic Compounds

Compounds	Peak ratio[a]	Sensory test[b]
Vanillin (1)	0.15	○
Phenol (2)	0.59	×
Benzaldehyde (3)	0.94	○
Anisole (4)	0.97	○
p-Hydroxybenzaldehyde (5)	0.11	○
m-Hydroxybenzaldehyde (6)	0.38	×
o-Methoxyphenol (7)	0.65	○
m-Methoxybenzaldehyde (8)	0.95	×
o-Vanillin (9)	0.13	○
iso-Vanillin (10)	0.28	×
Acetovanillone (11)	0.15	×
Syringic aldehyde (12)	0.12	○
Piperonal (13)	0.91	○
Eugenol (14)	0.73	×
iso-Eugenol (15)	0.57	×

[a]Peak ratio expressed as relative peak areas of the samples to that of initial TMA solution.
[b]○: Effective on pair test; ×: not effective.
Source: Ref. 15.

soup and reported that the aromas were suppressed by adding bay leaves or cloves [17,18]. They also evaluated the aromas generated by chicken soup containing thyme, thyme and rosemary, or thyme and clove [19]. The aromas generated by the soups containing two kinds of spice were found to be simply mixtures of the flavor compounds of each spice, proving that there were no chemical reactions between the different flavor compounds.

Recently, research on the masking of body odors including methyl mercaptan has attracted attention. Nutrients from food are broken down by digestive enzymes in the stomach and intestines and are then absorbed through the intestinal wall into the blood. After absorption the residues are sent to the colon and are then changed to feces. Ingested nutrients are decomposed by intestinal bacteria or enzymes in the gastrointestinal organs, and some of these decomposed nutrients contain bad smelling components. Some mercaptane compounds, hydrogen sulfide, and indole compounds, all of which are generated in the intestines, are absorbed into the blood through the intestinal wall and then sent to the liver. If the working of the liver is weakened because of a lesion or aging, it cannot properly decompose such substances so that they accumulate, resulting in bad breath and body odor. It is said that a high concentration of ammonia in the blood may be

TABLE 3.20 Change in Fishy Smell and Volatile Amines in Mackerel with Various Spices

Spice	Amount added (g)	Treatment	pH	Fishy smell	DMA-N (mg/100 g)	TMA-N (mg/100 g)	TMAO-N (mg/100 g)
Control		A	5.9	4	0.22	1.10	6.36
		B	5.9	4	0.33	1.79	5.94
Pepper	2	A	6.0	0	0.32	1.41	7.31
		B	6.0	0	0.32	1.38	6.05
Bay leaves	2	A	5.8	0	0.24	1.05	6.43
		B	5.8	0	0.33	1.35	5.67
Sage	2	A	6.0	0	0.22	1.26	6.85
		B	6.0	0	0.32	1.48	5.60
Japanese pepper	2	A	5.8	0	0.33	1.00	7.09
		B	5.7	0	0.36	1.27	6.29
Mustard	2	A	6.0	1	0.20	1.23	5.67
		B	5.9	1	0.32	1.46	6.72
Horseradish	2	A	5.8	2	0.24	1.09	6.67
		B	5.9	2	0.38	1.37	5.38
Curry powder	2	A	6.1	1	0.29	1.18	7.57
		B	6.1	2	0.53	1.84	5.64
Instant coffee	2	A	5.8	1	0.38	1.42	6.46
		B	5.8	2	0.60	2.11	5.86
Garlic	10	A	6.0	1	0.25	1.49	6.03
		B	6.1	1	0.38	1.91	6.01
Ginger	10	A	5.9	1	0.24	1.04	5.36
		B	6.0	1	0.38	1.24	5.54
Onion	10	A	5.9	2	0.26	1.08	6.22
		B	6.1	3	0.37	1.89	5.57
Radish	10	A	5.9	3	0.27	1.10	6.38
		B	6	3	0.43	1.88	5.42

Treatment A: Homogenizing mackerel with each spice, followed by leaving for an hour at room temperature.
Treatment B: Homogenizing mackerel with each spice, followed by boiling for 5 minutes and leaving for an hour.
Fishy smell: 4 (strong smell) > 3 > 2 > 1 > 0 (no smell).
Source: Ref. 16.

related to certain disease such as cancer, arteriosclerosis, or even Alzheimer's disease.

Abe et al. [20] wrote about the deodorizing effects of a plant extract (mushroom) against both ammonia in blood and foul smelling compounds in feces such as methyl mercaptan. They observed a change in the amount of methyl mercaptan and ammonia in the blood, thought to be an indicator of bad odor, with gas chromatography. They added this plant extract to livestock feed and fed it orally to rats for 30 days. Figure 3.7 shows the average values of both the blood ammonia concentration and gas concentration in feces before and after the 30-days feeding period. The test results indicate that both ammonia and gas concentration in feces

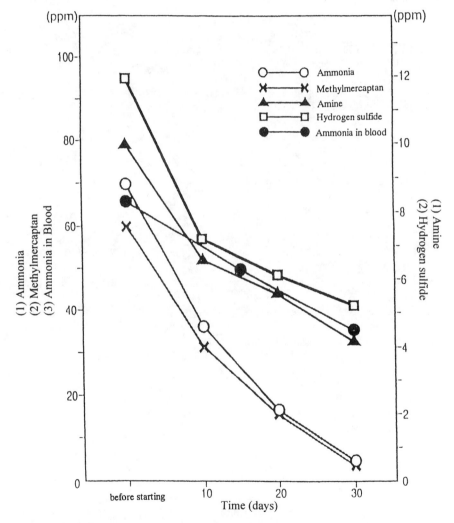

FIGURE 3.7 Change in body odor using plant extract. Average temperature: 29°C; humidity: 75%. (From Ref. 20.)

were remarkably decreased after feeding. They also suggested that the ammonia concentration in blood is correlated with that in feces.

As mentioned, some plant extracts were found to suppress body odor, and it is expected that the use of spices or spice extracts to deodorize/mask bad breath will attract attention in the future.

REFERENCES

1. Y. Kouchi, *Seikatu Eisei*, 38: 49 (1994).
2. K. Mori, *Food Chem.*, (11): 41 (1995).
3. H. Brogle, *Chem. Ind.*, 19: 385 (1982).
4. Y. Ito, H. Miura, and K. Miyaga, *Gyoniku Sausage Kyoukaisi*, 85: 24 (1962).
5. R. M. Cooper, *J. Geront.*, 14: 56 (1959).
6. S. Ohta, *Shokuhin chomiron*, Saiwai Shobo, Tokyo, 1976.
7. A. Goto, M. Komai, H. Suzuki, and Y. Furukawa, *Nihon Ajitonioi Gakkaishi*, 2: 339 (1995).
8. F. Tokita, M. Ishikawa, K. Shibuya, M. Koshimizu, and R. Abe, *Nippon Nogeikagaku Gakkaishi*, 58: 585 (1984).
9. J. M. Shewan, *Nature*, 143: 284 (1939).
10. W. J. Dyer and Y. A. Mouney, *J. Fish. Res. Bd. Canada*, 6: 359 (1945).
11. C. H. Castell, M. F. Greenugh, R. S. Rodgers, and A. S. MacFarlane, *J. Fish. Res. Bd. Canada*, 15: 701 (1958).
12. W. Shimizu and S. Hibiki, *Bull. Japan Soc. Sci. Fish.*, 23: 255 (1957).
13. T. Kikuchi, K. Hirai, and A. S. Sudarso, *Eiyo to Shokuryo*, 21: 253 (1968).
14. E. Niwa, M. Shibata, M. Nakamura, and M. Miyake, *Nippon Suisan Gakkaishi*, 37: 312 (1971).
15. K. Shiraishi, K. Imada, and H. Sugisawa, *Nippon Shokuhin Kogyo Gakkaishi*, 29: 426 (1982).
16. A Yoshida, S. Sasaki, and K. Okayama, *Seikatsu Eisei*, 27: 167 (1983).
17. F. Kawamura, T. Kawamura, K. Kato, M. Matsumoto, and A. Kobayashi, *Kasei Kagaku Gakkaishi*, 34: 387 (1983).
18. F. Kawamura, T. Hatanaka, M. Matsumoto, K. Kato, and A. Kobayashi, *Kasei Kagaku Gakkaishi*, 35: 681 (1984).
19. F. Kawamura, K. Kato, M. Matsumoto, T. Kawamura, and A. Kobayashi, *Kasei Kagaku Gakkaishi*, 35: 7 (1984).
20. J. Abe, K. Kanaya, and T. Toyoshima, *Tohoku Rinsho Eiseikensa Gakkai Koen Yoshisyu*, (36): (1995).

4

The Patterning Theory of Spice Use

I. PATTERNING FOR USING SPICES EFFICIENTLY

The use of spices in cooking varies greatly according to the area of the world in question. Sometimes spices not used in a native cuisine are introduced and used in a new way. For example, ginger is used in Japan in raw fish meal, such as finely chopped horse mackerel or lightly roasted bonito, while in Europe it is often used with pork. The combination of ginger and pork also occurs in Japanese cooking, such as in grilled pork with ginger. Sliced ginger may be added to stewed giblets for the purpose of eliminating the unpleasant smell of giblets.

The relationship of spices to cooking ingredients and techniques can be thought of as a kind of pattern [1]. The "patterning theory" of spice use considers the suitability of a spice and certain raw materials to be the result of a "synthesis" that occurs in the mouth. A preference for a specific spice is determined by individual judgment, and the suitability of any spice and any other ingredient is based on individual preference. Flavor preference is evaluated using the senses of taste and of smell. If a spice and an ingredient tasted together are well received, it follows that these two can be combined in cooking. Concrete examples of this application of the patterning theory are given in this chapter.

A. Patterning Synthesis in the Mouth

The Japanese confection *Daifuku*, containing sweet bean paste, is considered compatible in taste with green tea, and the two are often consumed together. It is, therefore, not surprising that products such as a sweet jellied bean paste containing both ingredients are available on the market. This patterning theory can be applied to Western and Chinese foods as well as to Japanese foods.

Similarly, if drinking coffee while eating a cake containing certain spices is pleasurable, it would make sense to mix coffee with those spices (Fig. 4.1). Along the same way of thinking, new types of spiced teas or drinks could be formulated.

B. Patterning the Suitability of Spices and Cooking Ingredients by Substituting Analogous Ingredients

Ground meat is a major ingredient in many foods. Nutmeg is often added to ground meat to deodorize or neutralize its meat smell. This use for nutmeg can be applied across the broad range of raw materials that have meat or meatlike flavors, such as milk and cream. Therefore, a spice used to enhance meat flavor can be used in a stew that contains milk or a cake to which fresh cream is added, although the timing or amount of spice may need to be adjusted (Fig. 4.2). For example, if pepper or ginger is suitable for use with meat, it should be possible to develop a kind of ice cream to which pepper or ginger is added.

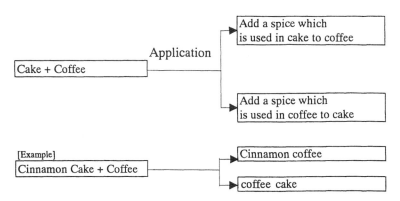

FIGURE 4.1 Patterning synthesis in the mouth.

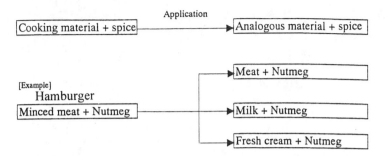

FIGURE 4.2 Patterning the suitability between spice and cooking ingredients.

C. Patterning the Suitability of Spices and Cooking Ingredients by Substituting Analogous Spices

Patterning can also be used in choosing a kind of spice for a specific raw ingredient. Spices with similar flavors tend to contain similar kinds of flavor compounds (Fig. 4.3). Patterning involves replacing one spice with others that have similar flavor characteristics.

D. Patterning the Suitability of Spices and Raw Materials by Substituting Spices from the Same Botanical Family

In an extension of the patterning theory used above, one spice belonging to the same botanical family can be substituted for another for the specific cooking ingredient because spices belonging to the same family tend to have similar series of flavor compounds and similar flavor characteristics. For example, onion, garlic, and leek all belong to the Liliaceae family. Their similar flavor characteristics

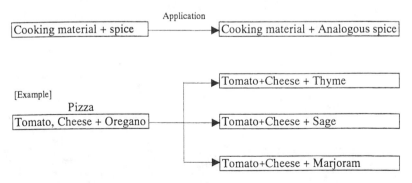

FIGURE 4.3 Patterning the suitability between spice and cooking ingredients.

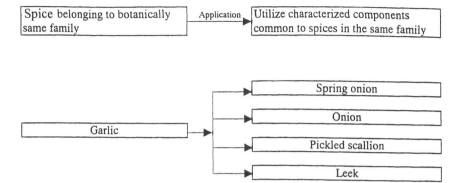

FIGURE 4.4 Patterning the suitability between spices and cooking ingredients by substituting spices belonging to the same botanical family.

enable them to be substituted for one another (Fig. 4.4). For example, garlic is usually used in *Gyoza*, a Chinese dish in which minced pork is wrapped in a small pancake and steamed or fried. Nira (*Allium ampleloprasum* L.), which belongs to the same botanical family as garlic, can be substituted for garlic in this dish.

E. Patterning the Suitability of Spices and Cooking Ingredients by Substituting Spices with Similar Pungency

The pungent spices shown in Figure 4.5 all belong to the Cruciferae family. All spices belonging to this family have similar pungent characteristics and can be used interchangeably.

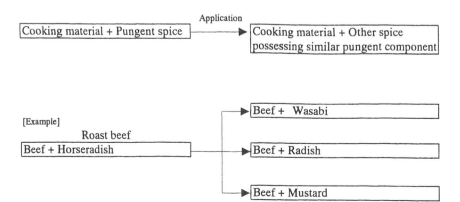

FIGURE 4.5 Patterning the suitability between spice and cooking ingredients by substituting spices with similar pungency.

For example, horseradish is often used to season roast beef. Radish, which belongs to the same family (Cruciferae) can be used instead if one wants to add a Japanese flavor accent. Soy sauce is also used to cook Japanese-flavored beef steak because its flavor is known to be compatible with that of radish. Horseradish powder is utilized as an imitation "wasabi" powder by adding a green color, but it also tends to be acceptable to the Japanese palate if Japanese mustard, which also belongs to the Cruciferae family, is added.

F. Patterning the Combination of Spices and Seasonings by Replacing Western with Japanese Seasoning

A common cooking technique involves the use of vinegar, wine, or spirits to season certain cooked foods. If a certain spice soaked in wine vinegar is found to be a good combination, for example, Japanese rice vinegar could also be used (Fig. 4.6). Japanese sake or shochu (Japanese spirits distilled from sweet potatoes) can be used in place of liqueur, which is another kind of alcoholic drink. There are many kinds of alcohol to which spices are added. Gin is usually used as the base of a cocktail, and several kinds of spices are added during the process of brewing gin. Gin or rum is often added to fruitcake, and spices can be added to make the flavor of fruitcake more characteristic. Red pepper alcohol, which is made by soaking red pepper in Japanese shochu, can be used as a new kind of seasoning. Table 4.1 presents a list of spices that can be used with various kinds of alcohol.

II. EVALUATING A SPICE'S SUITABILITY

When thinking about what spices to use in cooking, the name of the dish or the cooking method is one important factor to be considered. However, it is not necessarily true that a specific spice would be suitable only for some specific meals or specific ingredients. Any spice can be used in a variety of cooking applications.

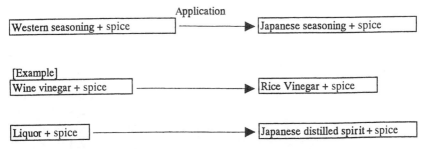

FIGURE 4.6 Patterning the combination of spices and seasonings by replacing Western with Japanese seasoning.

TABLE 4.1 Kinds of Spices Used in Various Alcoholic Beverages

Alcoholic beverage	Spices contained
Vermouth	Marjoram, sage, coriander, ginger, cardamom, cinnamon, clove, mace, peppermint, thyme, anise, juniper berry (vanilla)
Gin	Coriander, juniper berry
Aquavit	Anise, fennel, dill, caraway
Curacao	Cinnamon, clove, nutmeg, coriander
Kummel	Caraway, fennel, coriander
Anisette	Anise, fennel, nutmeg
Goldwasser	Caraway, coriander
Gancia	Cinnamon, cardamom, coriander, mint, fennel, clove, pepper
Creme de cumin	Cumin
Creme de cacao	Clove, mace (vanilla)
Creme de menthe	Peppermint
Peppermint schnapps	Peppermint

Choosing a spice for a certain food or cooking technique is often influenced by the preference of the inhabitants of the nation where the combination will be used. We have evaluated the suitability of certain spices to the foods of various nations, cooking ingredients, and cooking techniques, by checking a variety of combinations used in recipes in which many spices are used. Our evaluation method, called "Frequency Patterning Analysis," evaluates the suitability of each spice according to the above parameters.

A. Frequency Patterning Analysis

1. Theory behind the method

The suitability of a spice for any cooking ingredient or technique is determined in this analytical method based on a variety of combinations of spices with those ingredients or techniques in a variety of recipes in which various kinds of spices are used. The name "frequency patterning analysis" is used because the frequency of each of 40 spices is analyzed so that a pattern of spice use for each nation, each cooking ingredient, and each cooking technique can be seen.

2. The analytical method

The following steps are followed in performing this analysis:

1. Choose 40 spices that appear in recipes worldwide and analyze statistically their frequency of appearance according to nation, cooking ingredients, and cooking technique; then evaluate the suitability of each spice according to each of these factors.

The Patterning Theory of Spice Use

2. Divide cooking ingredients into vegetable and animal ingredients, then subdivide into nine categories: meats, fish, milk, cereals, potatoes, fruits, beans, seeds, and beverages. Count the number of times each spice appears under each ingredient. This number is the frequency of appearance of that specific spice.
3. Divide cooking techniques after adding spices into heat and nonheat cooking, then subdivide into eight categories: boiling, baking (including grilling), frying, deep-frying, steaming, pickling, dressing with sauce, and not cooking. Count the number of times that each spice appears in each category. That number is the frequency of appearance of that specific spice.
4. Divide types of cooking or meals into French, Italy, Germany, British, American (Western-style) and Southeast Asian, Chinese, Japanese, Indian, etc. (Eastern-style). Count the number of times each spice appears in each category. That number is the frequency of appearance of that specific spice.
5. Calculate the amount of each spice (percent value) for cooking ingredient, cooking technique, nation, respectively, based on frequency of appearance obtained for each case.
6. Calculate the suitable index of each spice for each of the categories—cooking ingredient, cooking technique, nation—based on the above calculated percentage of each spice.
7. Calculate the mean value of the spice from the values obtained above. A spice having a higher value than the calculated mean value is considered "suitable." Any spice having a much higher value is considered "very suitable."

3. Considerations during analysis

Points to consider during analysis are as follows:

1. Eliminate and try not to count items that are hard to interpret.
2. Do not count any spice more than once in the same category.
3. New data can always be added, and the reliability of the analysis is enhanced with the amount of data added.
4. When blended spices are used, count each spice separately.

4. Classification of spices

Frequency patterning analysis is based on 40 of 88 botanically classifiable spices commonly used worldwide (Table 4.2). If a different number or selection of spices is to be used in frequency patterning analysis, a new set of data needs to be gathered and analyzed.

TABLE 4.2 Botanical Classification of Spices

Angiospermae			
Dicotyledoneae			
Sympetalae			
	Tubiflorae	Boraginaceae	Borage
		Labiatae	Mint, marjoram, lavender, hyssop, thyme, basil, perilla, rosemary, oregano, savory, sage
		Solanaceae	Red pepper, paprika
		Pedaliaceae	Sesame
	Campanulatae	Comperaceae	Tarragon, wormwood, dandelion, tansy, japanese mugwort, endive, chicory, sunflower seed
	Rubiales	Rubiaceae	Gardenia
	Cucurbitales	Cucurbitaceae	Pumpkin, watermelon
Archichlamydeae			
	Piperales	Piperaceae	Pepper
	Urticales	Moraceae	Hemp, hop
	Polygonales	Polygonaceae	Water pepper, sorrel
	Ranales	Myristicaceae	Nutmeg, mace
		Lauraceae	Bay leaves, cinnamon
		Magnoliaceae	Star anise
	Rhoeadales	Cruciferae	Mustard, radish, cress, garden cress, horseradish, mustard green
		Papaveraceae	Poppy
		Capparidaceae	Caper
	Rosales	Roseceae	Rowan, almond, salad burnet
		Leguminosae	Fenugreek
	Geraniales	Tropaeolaceae	Nasturtium
		Erythroxylaceae	Coca
		Rutaceae	Japanese pepper, yuzu, rue
	Malvaceae	Malvaceae	Hibiscus
		Sterculiaceae	Cola
	Myrtiflorae	Myrtaceae	Allspice, clove
	Umbelliflorae	Umbelliferae	Parsley, celery, dill, cumin, anise, fennel, caraway, coriander, water dropwort, Japanese Honeywo, ajwain, chervil, angelica
Monocotyledoneae			
	Glumiflorae	Gramineae	Lemongrass
	Liliiflorae	Liliaceae	Garlic, onion, baker's garlic, chive, eshlot, asatsuk, lily, leek, tufted stone leek, welsh onion
		Iridaceae	Saffron
	Scitamineae	Zingiberaceae	Cardamom, ginger, mioga, turmeric, guinea grains
	Orchidales	Orchidaceae	Vanilla

5. Example of analysis

Figure 4.7 shows a recipe for white kidney bean soup, which can be used to explain the frequency patterning analytical procedure:

1. Onion, leek, pepper, parsley, celery, and bay leaves are the spices to be analyzed.
2. Chicken stock, pork, butter, carrot, kidney beans, and other ingredients are also used in this recipe and one point each is given to meat (chicken stock, pork), milk (butter), vegetable (carrot), and beans (kidney beans).
3. As explained above, a cooking technique is analyzed after adding a spice. In this case, one point each for onion and leek is allotted to "frying" because both spices are added before the frying process, and another point each for onion, leek, parsley, celery, and bay leaves is allotted to "boiling" because these spices are added as a bouquet garni before the boiling process.

The procedure to analyze the suitability of each spice is shown in Figure 4.8.

6. Result of analysis

Tables 4.3–4.5 show the results of analysis of cooking ingredients, cooking techniques, and nations, using 9000 recipes from around the world.

7. Utilization and application of the analytical results

Selecting a Suitable Spice According to Major Function It is possible to select a spice to fulfill a certain function based on the suitability values as determined above. As mentioned in the previous chapter, spices can be used to flavor, deodorize (mask), color, or add pungency. Each spice has a major function (A-value), which can be used to categorize spices selected from the principles of frequency patterning analysis.

Selecting a Suitable Spice According to General Function A spice may have subfunctions (B-value) in addition to its major function. For example, the major function of red pepper is to give pungency, but it also has the subfunction of acting as a colorant. Paprika, on the other hand, does not have any flavoring, deodorizing, or other function other than its coloring function. Table 4.6 shows spices classified according to major function (A) and subfunctions (B).

Selecting a Suitable Spice by Cross Reference Selecting spices can be made easier by referring to the indexes for several categories of spice suitability. For example, to find a spice suitable for boiling a meat in an Italian-style meal, cooking ingredients, cooking technique, and national indexes can be consulted.

B. National Trends in Spice Use

Spice use trends for individual countries obtained by frequency patterning analysis are arranged according to four functions: flavoring, deodorizing/masking,

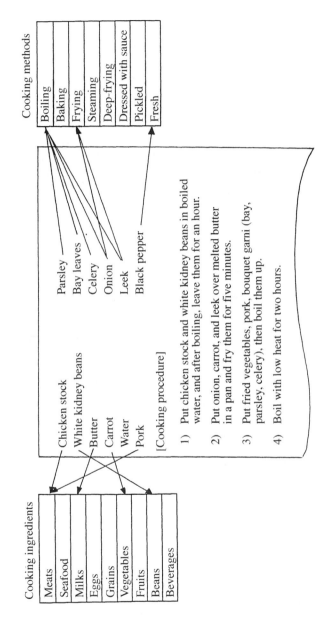

Figure 4.7 Recipe for white kidney bean soup.

The Patterning Theory of Spice Use

Spices	Frequency of appearance of each spice for each nation			The ratio each spice appearing			The suitable index of each spice		
	Japan	India	United States	Japan	India	United States	India	United States	TOTAL
01 Anise	0	12	63	0	1.4	5.2	14	52	100
02 Onion	24	45	79	5.1	5	6.5	10	13	100
03 Oregano	0	0	15	0	0	1.2	0	12	100
04 Allspice	0	0	20	0	0	1.6	0	16	100
32 Pepper	6	64	117	1.2	7.2	9.6	12	16	100
33 Horseradish	54	4	15	11.1	0.4	1.2	2	6	100
34 Marjoram	0	0	4	0	0	0.3	0	3	100
35 Mustard	13	17	23	2.7	1.9	1.9	12	12	100
36 Mint	0	10	38	0	1.1	3.1	11	32	100
37 Mace	0	29	2	0	3.3	0.2	56	3	100
38 Leek	68	0	16	14	0	1.3	0	3	100
39 Red pepper	39	107	98	8	12	8	15	10	100
40 Rosemary	0	0	15	0	0	1.2	0	12	100
TOTAL	485	890	1222	100	100	100	667	597	4000

FIGURE 4.8 Procedure for patterning analysis.

cooking, and pungency (Tables 4.7–4.10). Each spice is also arranged as to its use in Western cooking and Eastern cooking according to major function (Tables 4.11, 4.12).

When the incidence of spice use in Eastern cooking is compared with that in Western cooking, it is found that sesame seed, star anise, fenugreek, and cardamom are often used to flavor Eastern dishes, whereas tarragon, basil, allspice, marjoram, and vanilla are used more often in Western foods. This makes sense when the origins of the spices are taken into account, e.g., star anise from China, cardamom from India, fenugreek and sesame seed from the Middle or Near East, and allspice from South and Central America (although there are exceptions).

Rosemary, sage, oregano, thyme, and celery are commonly used in Western cooking for deodorizing purposes. Rosemary, sage, and oregano, all belonging to the Labiatae family, are routinely used in Mediterranean countries such as Italy and Greece. There are no great differences between Western and Eastern cooking as far as the use of spices as colorants or for pungency. One exception would be ginger, a spice of Eastern origin. Ginger is often used in Japanese and Chinese cooking for the purpose of masking the smell of seafood as well as to lend pungency, whereas in Western countries this spice is used most often in cakes to enhance the sweet flavor.

Basil, tarragon, dill, anise, paprika, mint, parsley, allspice, and pepper are spices suitable for American cooking. On the other hand, star anise and Japanese pepper, which are consumed often in China and Japan, are almost never used in American cooking. As for Indian cooking, fenugreek, cardamom, cumin, mace, turmeric, saffron, coriander, fennel, cinnamon, clove, red pepper, and nutmeg

TABLE 4.3 Spice Suitability According to Nation

Spice		JP	CH	SEA	IN	US	UK	GE	IT	FR
1.	Parsley		○	○	○	◉	◉	○	○	○
2.	Celery			○		○	○	○	◉	◉
3.	Cinnamon			○	◉	○	○	○		○
4.	Allspice					◉	◉	◉	○	
5.	Dill					○	◉		○	○
6.	Mint					○	◉	◉	○	
7.	Tarragon					◉		○		○
8.	Cumin				◉	◉	○			
9.	Marjoram							◉	○	○
10.	Star anise			◉	○					
11.	Basil					◉			○	
12.	Anise					○	◉			
13.	Mace					◉		○		
14.	Fennel					◉		◉		
15.	Sesame		◉	○						
16.	Vanilla								◉	
17.	Fenugreek					◉				
18.	Cardamom					◉				
19.	Garlic		○	○	○	○	○	○	○	○
20.	Onion		○		◉	○	◉	○	○	○
21.	Bay leaves					○				
22.	Clove					◉	○	○	○	○
23.	Nutmeg					○	○	◉	◉	○
24.	Leek		◉	○	○				○	○
25.	Thyme						○	○	○	○
26.	Rosemary						○	◉	○	◉
27.	Caraway				○	○		○	◉	
28.	Sage						○	◉		○
29.	Oregano						○			◉
30.	Savory								◉	
31.	Coriander			○		◉		○		
32.	Pepper			○	○	○	◉	○	○	○
33.	Red pepper		○	○	◉	○	○	○		
34.	Mustard		◉			○	○	○	○		○
35.	Ginger		○	◉	○	○	○			
36.	Horseradish		◉					○	○	
37.	Japanese pepper		◉	○						
38.	Paprika					○		◉	◉	
39.	Turmeric					○	◉		○	
40.	Saffron					○				◉

Suitability: ○ < ◉.
JP: Japan, CH: China, SEA: Southeast Asia, IN: India, UK: United Kingdom, US: United States, IT: Italy, FR: France, GE: Germany.

TABLE 4.4 Spice Suitability According to Ingredient

Spice		ME	SF	MI	EG	GR	VE	FR	BS	BE
1.	Parsley	O	O	O	O	O	O			⊙
2.	Celery	O	O	O	O	O	O			⊙
3.	Cinnamon	O		O		O	O	O	⊙	O
4.	Allspice	O	O	O	O	O	O	O		O
5.	Dill	O	O	O	O	O	O	O		O
6.	Mint	O	O	O	O	O	O	O	O	
7.	Tarragon	O	O	O	O	O	O	O	O	O
8.	Cumin	O		O		O	O	O	⊙	
9.	Marjoram	O		O	O	O	O	O	O	
10.	Star anise	O	O	O	O	O	O	O	O	O
11.	Basil	O	O	O		O	O	O	O	O
12.	Anise	O	⊙	O	O	O	O	O		
13.	Mace			O	O	O	O	O	O	O
14.	Fennel	O	⊙	O		O	O	O	O	O
15.	Sesame	O			O	O	O		O	O
16.	Vanilla			⊙	O	O		⊙		O
17.	Fenugreek		⊙							
18.	Cardamom	O		O				O	⊙	
19.	Garlic	⊙	O		O			O	O	
20.	Onion	O	O	O	O	O	O	O	O	
21.	Bay leaves	O	O	O	O	O	O	O		⊙
22.	Clove	O	O	O	O	O	O	O	O	O
23.	Nutmeg	O	O	O	O	O	O	O	O	O
24.	Leek		O	O	⊙	O	O		O	O
25.	Thyme	O	O	O	⊙	O	O			⊙
26.	Rosemary	⊙	O	O		O	O	O		O
27.	Caraway			⊙	O	⊙	O	O		O
28.	Sage	⊙	O	O	O	O	O	O	O	O
29.	Oregano	O	O	O	O	O	O	O	O	O
30.	Savory	O					O	O	⊙	
31.	Coriander						⊙			⊙
32.	Pepper	O	O	O	O	O	O	O	O	O
33.	Red pepper	O	O	O	O	O	O	O	O	
34.	Mustard	O	O	O	O	O	O	O	O	O
35.	Ginger	O	O	O	O	O	O	O	⊙	O
36.	Horseradish		⊙	O	O	O	O	O		O
37.	Japanese pepper	⊙	⊙		O	O	O		O	O
38.	Paprika	O	O	O	O	O	O	O	O	O
39.	Turmeric	O	O	O	O	O	O	O	⊙	
40.	Saffron	O	⊙	O	O	O	O	O	O	

Suitability: O < ⊙.
ME: meat, SF: seafood, MI: milk, EG: egg, GR: grains, VE: vegetables, FR: fruits, BS: bean and seeds, BE: beverages.

TABLE 4.5 Spice Suitability According to Cooking Technique

Spice	SI	BK	FR	ST	DF	DR	PK	FS
1. Parsley	○				○	⊙	○	⊙
2. Celery	⊙	○	○		○	⊙	⊙	⊙
3. Cinnamon	○	○	○	⊙	○	○	○	○
4. Allspice	○	○	○	⊙	○	○	⊙	○
5. Dill	○	○	○	⊙	○	○	⊙	○
6. Mint	○					○	○	⊙
7. Tarragon	○					○	⊙	⊙
8. Cumin	○	○	○	○	⊙	○	○	○
9. Marjoram	⊙	○		○	○	○	○	○
10. Star anise	⊙						○	⊙
11. Basil	⊙					○	○	⊙
12. Anise		○					⊙	○
13. Mace	○	○	⊙	○	○		⊙	○
14. Fennel	⊙	⊙	○	○	○	○	○	⊙
15. Sesame		⊙	⊙	○	⊙			
16. Vanilla	○	⊙		⊙	○	○	○	⊙
17. Fenugreek	○	⊙	○	○	○			
18. Cardamom	○	○	○	⊙	○	○	○	⊙
19. Garlic	○	⊙	⊙	○	○	○	○	○
20. Onion	⊙	⊙	⊙	○	○	○	○	○
21. Bay leaves	⊙	⊙	○	○	○	○		
22. Clove	⊙	○	⊙	○	○	○		
23. Nutmeg	○	⊙	⊙	⊙	○		⊙	○
24. Leek	○	⊙	⊙	○	⊙	○	○	○
25. Thyme	⊙	○	○	○	○			
26. Rosemary	⊙	⊙	⊙	○	⊙			
27. Caraway	○	○	○	○	○	⊙	○	○
28. Sage	○	⊙	○	⊙	⊙	○		
29. Oregano	○	⊙	⊙	○	⊙		○	○
30. Savory	○	○	○	○	○	○	○	○
31. Coriander	○	○	○	○	○	○	⊙	⊙
32. Pepper	○	⊙	○	○	○	○	○	○
33. Red pepper	○	○	○	○	○	⊙	⊙	○
34. Mustard						⊙	⊙	⊙
35. Ginger	○	○	⊙	⊙	○	○	○	○
36. Horseradish						⊙	⊙	⊙
37. Japanese pepper						⊙	⊙	⊙
38. Paprika	○	○	○	○	○	○	○	○
39. Turmeric	○	○	⊙	⊙	○	⊙	○	
40. Saffron	⊙							

Suitability: ○ < ⊙.
SI: simmered (boiled) food, BK: baked food, FR: fried food, ST: steamed food, DF: deep-fried food, DR: food dressed with sauce, PK: pickled food, FS: fresh food.

The Patterning Theory of Spice Use

TABLE 4.6 Basic Working of Spices

Basic function	A (major function)	B (subfunction)
Flavoring	Parsley, cinnamon, allspice, dill, mint, tarragon, cumin, marjoram, star anise, basil, anise, mace, nutmeg, fennel, sesame, vanilla, fenugreek, cardamom, celery	Garlic, onion, bay leaves, clove, thyme, rosemary, caraway, sage, savory, coriander, pepper, oregano, horseradish, Japanese pepper, saffron, ginger, leek, mustard
Deodorizing/ Masking	Garlic, savory, bay leaves, clove, leek, thyme, rosemary, caraway, sage, oregano, onion, coriander	Parsley, pepper, allspice, mint, tarragon, cumin, star anise, mace, fennel, sesame, cardamom, mustard, cinnamon, vanilla, horseradish, Japanese pepper, nutmeg, ginger
Pungency	Japanese pepper, mustard, ginger, horseradish, red pepper, pepper	Red pepper
Coloring	Paprika, turmeric, saffron	

TABLE 4.7 National Trends for Spice Use—Flavor

	Eastern cooking				Western cooking				
Spice	JP	CH	SEA	IN	US	UK	GE	IT	FR
Parsley		O	O	O	⊙	⊙	O	O	O
Celery		O			O	O	O	⊙	⊙
Cinnamon			O	⊙	O	O	O		O
Allspice					⊙	⊙	⊙	O	
Dill				O	⊙		O		O
Mint				O	⊙	⊙		O	
Tarragon					⊙		O		O
Cumin				⊙	⊙	O			
Marjoram						⊙	O	O	
Star anise		⊙	O						
Basil					⊙			O	
Anise				O	⊙				
Mace				⊙		O			
Fennel				⊙		⊙			
Sesame	⊙	O							
Vanilla								⊙	
Fenugreek				⊙					
Cardamom				⊙					

Suitability: O < ⊙.
Abbreviations as for Table 4.3.

TABLE 4.8 National Trends for Spice Use—Deodorizing/Masking Effect

	Eastern cooking				Western cooking				
Spice	JP	CH	SEA	IN	US	UK	GE	IT	FR
Garlic	O	O	O	O	O	O	O	O	O
Onion	O		⊙	O	⊙	O	O	O	O
Bay leaves			O		O	O	O	O	⊙
Clove				⊙	O	O	O	O	O
Nutmeg				O	O	⊙	⊙	O	O
Leek	⊙	O	O					O	O
Thyme					O	O	O	O	⊙
Rosemary					O	⊙	O	⊙	O
Caraway			O	O	O	⊙			
Sage					O	⊙		O	
Oregano					O			⊙	
Savory						⊙			
Coriander		O		⊙			O		

Suitability: O < ⊙.
Abbreviations as for Table 4.3.

were found to be suitable. These spices are cultivated and processed mainly in India. The spice mixture garam masala, widely used in Indian cooking, consists of a blend of five or more of the above spices.

Sesame, leek, garlic, onion, horseradish, Japanese pepper, mustard, red pepper, and ginger were found to be very suitable for Japanese cooking. Sesame, Japanese pepper, red pepper, and ginger are often used for making a Japanese-

TABLE 4.9 National Trends for Spice Use—Pungency

	Eastern cooking				Western cooking				
Spice	JP	CH	SEA	IN	US	UK	GE	IT	FR
Pepper		O	O	O	⊙	O	O	O	O
Red pepper	O	O	⊙	⊙	O	O			
Mustard	⊙			O	O	O	O		O
Ginger	O	⊙	O	O		O			
Horseradish	⊙					O	O		
Japanese pepper	⊙	O							

Suitability: O < ⊙.
Abbreviations as for Table 4.3.

The Patterning Theory of Spice Use

TABLE 4.10 National Trends for Spice Use—Colorants

	Eastern cooking				Western cooking				
Spice	JP	CH	SEA	IN	US	UK	GE	IT	FR
Paprika			O		⊙		⊙		
Turmeric			O	⊙		O			
Saffron				⊙				⊙	

Suitability: O < ⊙.
Abbreviations as for Table 4.3.

flavored spice mixture. Star anise, ginger, sesame, and coriander show very high suitability for Chinese cooking. The Chinese five-spice mixture *uoshanfen* includes star anise, Japanese pepper, and others. There are similar cases for other countries: red pepper and cumin for Southeast Asian cooking; marjoram, sage, mint, allspice, rosemary, parsley, fennel, and nutmeg for British cooking; savory, caraway, nutmeg, and paprika for German cooking; vanilla, oregano, rosemary, celery, and saffron for Italian cooking; and thyme, bay leaves, and celery for French cooking (Tables 4.13–4.21).

These tables can be used to expand spice suitability for further applications. Forty kinds of spices are classified according to the major function of each spice. But some of these spices have subfunctions that play important roles. The case of spice suitability to American cooking is shown in Table 4.22. The major function of red pepper is to lend pungency to a dish, but it may also be used to fulfill its subfunction as a colorant.

C. Possible New Uses for Spices

A nation's typical cuisine tends to change gradually, being affected by regional cooking trends around the world. In Japan, a type of regional cooking called "ethnic meals" has become popular, but the flavors are usually altered to taste

TABLE 4.11 Spices Used in Eastern Cooking

Function	Spices
Flavoring	Parsley, cinnamon, dill, mint, cumin, star anise, anise, mace, nutmeg, fennel, sesame, fenugreek, cardamom, celery
Deodorizing/ Masking	Garlic, bay leaves, clove, leek, caraway, onion, coriander
Pungency	Japanese pepper, mustard, ginger, horseradish, red pepper, pepper
Coloring	Paprika, turmeric, saffron

TABLE 4.12 Spices Used in Western Cooking

Function	Spices
Flavoring	Parsley, cinnamon, allspice, dill, mint, tarragon, cumin, marjoram, basil, anise, mace, nutmeg, fennel, vanilla, celery
Deodorizing/Masking	Garlic, savory, bay leaves, clove, leek, thyme, rosemary, caraway, sage, oregano, onion, coriander
Pungency	Mustard, ginger, horseradish, red pepper, pepper
Coloring	Paprika, turmeric, saffron

TABLE 4.13 Spices Used in Japanese Cooking

Function	Spices
Flavoring	Sesame
Deodorizing/Masking	Garlic, leek, onion
Pungency	Japanese pepper, mustard, ginger, horseradish, red pepper
Coloring	

TABLE 4.14 Spices Used in Chinese Cooking

Function	Spices
Flavoring	Parsley, star anise, sesame
Deodorizing/Masking	Garlic, leek, coriander
Pungency	Japanese pepper, ginger, red pepper, pepper
Coloring	

TABLE 4.15 Spices Used in Indian Cooking

Function	Spices
Flavoring	Parsley, cinnamon, dill, mint, cumin, anise, mace, nutmeg, fennel, fenugreek, cardamom
Deodorizing/Masking	Garlic, clove, caraway, onion, coriander
Pungency	Mustard, ginger, red pepper, pepper
Coloring	Turmeric, saffron

Table 4.16 Spices Used in Southeast Asian Cooking

Function	Spices
Flavoring	Parsley, cinnamon, cumin, star anise, celery
Deodorizing/Masking	Garlic, bay leaves, leek, caraway, onion
Pungency	Ginger, red pepper, pepper
Coloring	Paprika, turmeric

Table 4.17 Spices Used in the United States

Function	Spices
Flavoring	Parsley, cinnamon, allspice, dill, mint, tarragon, cumin, basil, anise, nutmeg, celery
Deodorizing/Masking	Garlic, bay leaves, clove, thyme, rosemary, sage, oregano, onion
Pungency	Mustard, red pepper, pepper
Coloring	Paprika

Table 4.18 Spices Used in the United Kingdom

Function	Spices
Flavoring	Parsley, cinnamon, allspice, mint, marjoram, mace, nutmeg, fennel, celery
Deodorizing/Masking	Garlic, bay leaves, clove, thyme, rosemary, caraway, sage, onion
Pungency	Mustard, ginger, horseradish, red pepper
Coloring	Turmeric

Table 4.19 Spices Used in German Cooking

Function	Spices
Flavoring	Parsley, cinnamon, allspice, dill, tarragon, marjoram, nutmeg, celery
Deodorizing/Masking	Garlic, savory, bay leaves, clove, thyme, rosemary, caraway, onion, coriander
Pungency	Mustard, horseradish, pepper
Coloring	Paprika

TABLE 4.20 Spices Used in Italian Cooking

Function	Spices
Flavoring	Parsley, allspice, mint, marjoram, basil, nutmeg, vanilla, celery
Deodorizing/Masking	Garlic, bay leaves, clove, leek, thyme, rosemary, sage, oregano, onion
Pungency	Pepper
Coloring	Saffron

more "Japanese." In Western cooking, there are cases in which flavors characteristic of French food are adapted to another nation's cuisine. Creole cuisine, found in and around New Orleans, is basically a kind of French cooking, but American flavor characteristics are added using chives, green pepper, and poppy seed, all of which are routinely used in American cooking. Chinoise cuisine is a kind of Chinese-French cooking in which some typical Chinese spices like ginger, coriander, Japanese pepper, and Chinese five-spice mixture are added to basic French food. TexMex cooking is another example of a flavor mixture of two nations in which anise, coriander, cumin, and chives, often found in Southwestern food, are used in addition to spices normally used in Mexico.

D. Trends in Spice Use According to Cooking Ingredient

Spice trends according to cooking ingredients obtained by frequency patterning analysis are arranged in the Tables 4.23–4.26 according to flavoring, deodorizing, coloring, and pungency functions. Each spice is also rated as to its major function in combination with either animal or plant materials (Tables 4.27, 4.28).

There is only a small difference between the suitability of spices for animal as opposed to plant materials. Exceptions include fenugreek, in many cases used only for animal materials for its flavoring function, and coriander, used specifically for animal materials for deodorizing purposes. For the purpose of coloring and pungency, no difference is found.

TABLE 4.21 Spices Used in French Cooking

Function	Spices
Flavoring	Parsley, cinnamon, dill, tarragon, nutmeg, celery
Deodorizing/Masking	Garlic, bay leaves, clove, leek, thyme, rosemary, onion
Pungency	Mustard, pepper
Coloring	

The Patterning Theory of Spice Use

TABLE 4.22 Spices Used (including subfunctions) in the United States

Function	Spices
Flavoring	Parsley, cinnamon, allspice, dill, mint, tarragon, cumin, basil, anise, nutmeg, celery, onion, rosemary, garlic, sage, oregano, thyme, clove, bay leaves, mustard, pepper
Deodorizing/ Masking	Garlic, bay leaves, clove, thyme, rosemary, sage, oregano, onion, tarragon, parsley, mint, allspice, celery, nutmeg, mustard, cinnamon, pepper, cumin
Pungency	Mustard, red pepper, pepper
Coloring	Paprika, red pepper

TABLE 4.23 Spices Suitable for Flavoring Animal and Plant Foods

Spice	Animal material				Plant material				
	ME	SF	MI	EG	GR	VE	FR	BS	BE
Parsley	O	O	O	O	O	O			⊙
Celery	O	O	O	O	O	O			⊙
Cinnamon	O		O		O	O	O	⊙	O
Allspice	O	O	O	O	O	O	O		O
Dill	O	O	O	O	O	O	O		O
Mint	O	O	O	O	O	O	O	O	
Tarragon	O	O	O	O	O	O	O	O	O
Cumin	O		O		O	O	O	⊙	
Marjoram	O		O	O	O	O	O	O	
Star anise	O	O	O	O	O	O	O	O	O
Basil	O	O	O		O	O	O	O	O
Anise	O	⊙	O	O	O	O	O		
Mace			O	O	O	O	O	O	O
Fennel	O	⊙	O			O	O	O	O
Sesame	O			O	O	O		O	O
Vanilla			⊙	O	O		⊙		O
Fenugreek		⊙							
Cardamom	O		O				O	⊙	

Suitability: O < ⊙.
Abbreviations as for Table 4.4.

TABLE 4.24 Spices Suitable for Deodorizing Animal and Plant Foods

Spice	Animal material				Plant material				
	ME	SF	MI	EG	GR	VE	FR	BS	BE
Garlic	⊙	○		○		○	○	○	
Onion	○	○	○	○	○	○	○	○	
Bay leaves	○	○	○	○	○	○	○		⊙
Clove	○	○	○	○	○	○	○	○	○
Nutmeg	○	○	○	○	○	○	○	○	○
Leek		○	○	⊙	○	○		○	○
Thyme	○	○	○	⊙	○	○			⊙
Rosemary	⊙	○	○		○	○	○		○
Caraway			⊙	○	⊙	○	○		○
Sage	⊙	○	○	○	○	○	○	○	○
Oregano	○	○	○	○	○	○	○	○	○
Savory	○					○	○	⊙	
Coriander					⊙				⊙

Suitability: ○ < ⊙.
Abbreviations as for Table 4.4.

Garlic, rosemary, sage, and Japanese pepper are found to have high suitability to meat. Garlic is widely used for many kinds of meat such as beef, pork, lamb, and chicken, meaning that this spice has a high suitability for most kinds of meats. Most spices found to be suitable to meat have a deodorizing function as well. Japanese peper also shows high suitability to meat as a flavoring spice. It is blended with salt in many fried meat dishes in Eastern nations.

TABLE 4.25 Spices Suitable for Adding Pungency to Animal and Plant Foods

Spice	Animal material				Plant material				
	ME	SF	MI	EG	GR	VE	FR	BS	BE
Pepper	○	○	○	○	○	○	○	○	○
Red pepper	○	○	○	○	○	○	○	○	
Mustard	○	○	○	○	○	○	○	○	○
Ginger	○	○	○	○	○	○	○	⊙	○
Horseradish		⊙	○	○	○	○	○		○
Japanese pepper	⊙	⊙			○	○	○		○

Suitability: ○ < ⊙.
Abbreviations as for Table 4.4.

The Patterning Theory of Spice Use

TABLE 4.26 Spices Suitable for Coloring Animal and Plant Foods

	Animal material				Plant material				
Spice	ME	SF	MI	EG	GR	VE	FR	BS	BE
Paprika	○	○	○	○	○	○	○	○	○
Turmeric	○	○	○	○	○	○	○	⊙	
Saffron	○	⊙	○	○	○	○	○	○	

Suitability: ○ < ⊙.
Abbreviations as for Table 4.4.

As for seafood, fenugreek, Japanese pepper, horseradish, anise, fennel, and saffron show high suitability. Fenugreek is an essential ingredient of curry powder or garam masala in Indian cooking. Japanese pepper and horseradish are routinely used for Japanese seafood dishes. Fennel is often used in fish dishes in Western nations. It is also used in China to prevent people from becoming ill after eating fish. Saffron is frequently used to cook bouillabaisse, risotto, pilaf, and paella, in which many kinds of seafood are used.

Caraway and vanilla are found to be suitable for combination with milk (i.e., fresh cream, cheese, yogurt, and milk). Vanilla is essential for making ice cream containing fresh cream, and caraway seed is mixed with many kinds of cheese products. Thyme and leeks are found to have high suitability to eggs. Both are known to have a deodorizing function, and are used to mask the typical egg smell, e.g., when cooking omelets.

As for suitability to plant materials, most spices were evaluated to be suitable for grain, with caraway showing especially high suitability for grains. This spice is widely used in Europe, especially in Germany, in breads, cookies, and cakes. Most spices are found to be suitable for vegetables, with coriander being evaluated as most suitable. The frequency of use of this spice is higher in

TABLE 4.27 Spices Used in Foods Containing Animal Ingredients

Function	Spices
Flavoring	Parsley, cinnamon, allspice, dill, mint, tarragon, cumin, marjoram, star anise, basil, anise, mace, nutmeg, fennel, sesame seed, fenugreek, cardamom, celery
Deodorizing/ Masking	Garlic, savory, bay leaves, clove, leek, thyme, rosemary, caraway, sage, oregano, onion
Pungency	Japanese pepper, mustard, ginger, horseradish, red pepper, pepper
Coloring	Paprika, turmeric, saffron

TABLE 4.28 Spices Used in Foods Containing Plant Ingredients

Function	Spices
Flavoring	Parsley, cinnamon, allspice, dill, mint, tarragon, cumin, marjoram, star anise, basil, anise, mace, nutmeg, fennel, sesame seed, vanilla, cardamom, celery
Deodorizing/Masking	Garlic, savory, bay leaves, clove, leek, thyme, rosemary, caraway, sage, oregano, onion, coriander
Pungency	Japanese pepper, mustard, ginger, horseradish, red pepper, pepper
Coloring	Paprika, turmeric, saffron

Eastern nations than in other areas, and it is especially routinely used in the cooking of India and China. In Eastern nations, the fresh leaves of this spice (which is called "shiang cai" in China) are often used in salads or other vegetable dishes. A relatively large number of spices are evaluated to be suitable to fruits, but only a few kinds show very high suitability. Vanilla is found to be most suitable because many fruits are utilized as ingredients of desserts in which vanilla is frequently used. Cardamom, savory, cumin, turmeric, and ginger show high suitability for beans. All of these spices (except savory) are ingredients mainly used in curry powder, which is often cooked with some kind of beans. These spices do not have a deodorizing function, and it is necessary to use another spice with a deodorizing/masking function when the curry dish contains meats or seafood. Savory is used in bean dishes in German and other Western cooking. Bay leaves, celery, parsley, coriander, and thyme are found to show high suitability for beverages. Celery and parsley are common constituents of vegetable juice. Both spices perform a deodorizing function or provide a "hidden flavor"—they are not expected to perform a flavoring function. Bay leaves, coriander, and thyme are expected to perform a deodorizing/masking function. Spice suitability summarized according to major function is shown in Tables 4.29–4.37. Table 4.38 summarizes spice suitability for meat, including subfunction as well as major function.

TABLE 4.29 Spices Used in Meat Dishes

Function	Spices
Flavoring	Parsley, cinnamon, allspice, dill, mint, tarragon, cumin, marjoram, star anise, basil, anise, nutmeg, fennel, sesame, cardamom, celery
Deodorizing/Masking	Garlic, savory, bay leaves, clove, thyme, rosemary, sage, oregano, onion
Pungency	Japanese pepper, mustard, ginger, red pepper, pepper
Coloring	Paprika, turmeric, saffron

The Patterning Theory of Spice Use

TABLE 4.30 Spices Used in Seafood Dishes

Function	Spices
Flavoring	Parsley, allspice, dill, mint, tarragon, star anise, basil, anise, nutmeg, fennel, fenugreek, celery
Deodorizing/Masking	Garlic, bay leaves, clove, leek, thyme, rosemary, sage, oregano, onion
Pungency	Japanese pepper, mustard, ginger, horseradish, red pepper, pepper
Coloring	Paprika, turmeric, saffron

TABLE 4.31 Spices Suitable for Combining with Milk Products

Function	Spices
Flavoring	Parsley, cinnamon, allspice, dill, mint, tarragon, cumin, marjoram, star anise, basil, anise, mace, nutmeg, fennel, vanilla, cardamom, celery
Deodorizing/Masking	Bay leaves, clove, leek, thyme, rosemary, caraway, sage, oregano, onion
Pungency	Mustard, ginger, horseradish, red pepper, pepper
Coloring	Paprika, turmeric, saffron

TABLE 4.32 Spices Suitable for Combining with Eggs

Function	Spices
Flavoring	Parsley, allspice, dill, mint, tarragon, marjoram, star anise, anise, mace, nutmeg, sesame, vanilla, celery
Deodorizing/Masking	Garlic, bay leaves, clove, leek, thyme, caraway, sage, oregano, onion
Pungency	Japanese pepper, mustard, ginger, horseradish, red pepper, pepper
Coloring	Paprika, turmeric, saffron

TABLE 4.33 Spices Suitable for Combining with Grains

Function	Spices
Flavoring	Parsley, cinnamon, allspice, dill, mint, tarragon, cumin, marjoram, star anise, basil, anise, mace, nutmeg, fennel, sesame seed, vanilla, celery
Deodorizing/Masking	Bay leaves, clove, leek, thyme, rosemary, caraway, sage, oregano, onion
Pungency	Japanese pepper, mustard, ginger, horseradish, red pepper, pepper
Coloring	Paprika, turmeric, saffron

TABLE 4.34 Spices Suitable for Cooking with Vegetables

Function	Spices
Flavoring	Parsley, cinnamon, allspice, dill, mint, tarragon, cumin, marjoram, star anise, basil, anise, mace, nutmeg, fennel, sesame seed, celery
Deodorizing/ Masking	Garlic, savory, bay leaves, clove, leek, thyme, rosemary, caraway, sage, oregano, onion, coriander
Pungency	Japanese pepper, mustard, ginger, horseradish, red pepper, pepper
Coloring	Paprika, turmeric, saffron

TABLE 4.35 Spices Suitable for Cooking with Fruits

Function	Spices
Flavoring	Cinnamon, allspice, dill, mint, tarragon, cumin, marjoram, star anise, basil, anise, mace, nutmeg, fennel, vanilla, cardamom
Deodorizing/ Masking	Garlic, savory, bay leaves, clove, rosemary, caraway, sage, oregano, onion
Pungency	Mustard, ginger, horseradish, red pepper, pepper
Coloring	Paprika, turmeric, saffron

TABLE 4.36 Spices Suitable for Cooking with Beans and Seeds

Function	Spices
Flavoring	Cinnamon, mint, tarragon, cumin, marjoram, star anise, basil, mace, nutmeg, fennel, sesame, cardamom
Deodorizing/ Masking	Garlic, savory, clove, leek, sage, oregano, onion
Pungency	Japanese pepper, mustard, ginger, red pepper, pepper
Coloring	Paprika, turmeric, saffron

TABLE 4.37 Spices Suitable for Using in Beverages

Function	Spices
Flavoring	Parsley, cinnamon, allspice, dill, tarragon, star anise, basil, mace, nutmeg, fennel, sesame, vanilla, celery
Deodorizing/ Masking	Bay leaves, clove, leek, thyme, rosemary, caraway, sage, oregano, coriander
Pungency	Japanese pepper, mustard, ginger, horseradish, pepper
Coloring	Paprika

TABLE 4.38 Spices Suitable (including subfunctions) for Cooking Meat

Function	Spices
Flavoring	Parsley, cinnamon, allspice, dill, mint, tarragon, cumin, marjoram, star anise, basil, anise, nutmeg, fennel, sesame seed, cardamom, celery, onion, ginger, saffron, bay leaves, rosemary, clove, sage, oregano, garlic, Japanese pepper, pepper, mustard, thyme
Deodorizing/ Masking	Garlic, savory, bay leaves, clove, thyme, rosemary, sage, oregano, onion, sesame, tarragon, star anise, allspice, Japanese pepper, ginger, fennel, pepper, mustard, cardamom, nutmeg, cinnamon, cumin, celery, mint, parsley
Pungency	Japanese pepper, mustard, ginger, red pepper, pepper
Coloring	Paprika, turmeric, saffron, red pepper

E. Trends in Spice Use According to Cooking Technique

Spice trends according to cooking technique obtained by frequency patterning analysis are shown in Tables 4.39–4.42. Each spice is listed for both cooking with and without heat according to the major function of the spice (Tables 4.43, 4.44). Of spices with a flavoring function, fenugreek, sesame seed, and oregano can be used for cooking with heat. Fenugreek and sesame seed contain nonvolatile oils that tend to change upon heating so that their typical flavors are enhanced. Oregano, which differs from other Labiatae spices (e.g., basil and marjoram) in its flavor characteristics, possesses a "dull" flavor that can be weakened by the heating process. Among spices that have a deodorizing/masking function, rosemary, thyme, and star anise are suitable for cooking with heat because the heating process weakens the undesirable flavors of these spices. Most pungent spices such as mustard, horseradish, and Japanese pepper are more suitable for cooking without heat. Both mustard and horseradish achieve their pungency by the action of an enzyme, which is easily deactivated by the heating process. The typical flavor compounds contained in Japanese pepper tend to volatilize upon heating. Among spices that have a coloring function, saffron shows great suitability for cooking with heat. Since the coloring compounds of saffron are soluble in water, it is particularly effective to use this spice for boiling dishes, such as those containing rice (pilaf, paella).

Certain spices are well suited for use with particular methods of cooking with heat. Saffron, star anise, celery, marjoram, bay leaves, thyme, basil, fennel, onion, clove, and rosemary show high suitability for dishes that are boiled. Most of these have either a deodorizing or a flavoring function. Sesame seed, fenugreek, vanilla, rosemary, fennel, garlic, bay leaves, sage, onion, nutmeg, leek, oregano, and pepper are especially suitable for use in baked foods. It is well known that

TABLE 4.39 Suitability of Spices Used for Flavoring for Different Cooking Methods

Spice	Heating					Nonheating		
	SI	BK	FR	ST	DF	DR	PK	FS
Parsley	○				○	⊙	○	⊙
Celery	⊙	○	○		○	⊙	⊙	⊙
Cinnamon	○	○	○	⊙	○	○	○	○
Allspice	○	○	○	⊙	○	○	⊙	○
Dill	○	○	○	⊙	○	○	⊙	○
Mint	○					○	○	⊙
Tarragon	○					○	⊙	⊙
Cumin	○	○	○	○	⊙	○	○	○
Marjoram	⊙	○		○	○	○	○	○
Star anise	⊙						○	⊙
Basil	⊙					○	○	⊙
Anise		○					⊙	○
Mace	○	○	⊙	○	○		⊙	○
Fennel	⊙	⊙	○	○	○	○	○	⊙
Sesame		⊙	⊙	○	⊙			
Vanilla	○	⊙		⊙	○	○	○	⊙
Fenugreek	○	⊙	○	○	○			
Cardamom	○	○	○	⊙	○	○	○	⊙

Suitability: ○ < ⊙.
Abbreviations as for Table 4.5.

sesame seed produces appetizing flavor upon baking. Sesame seed, clove, thyme, garlic, onion, nutmeg, rosemary, mace, leek, oregano, ginger, and turmeric show high suitability for use in fried foods. (Interestingly, most of these also show high suitability for baked foods.) Cinnamon, allspice, turmeric, vanilla, thyme, dill, ginger, cardamom, and sage show high suitability for steamed foods. Most of these spices are used for making confectionery and are characterized by having relatively mild flavors. Cinnamon, dill, and cardamom are also suitable for non-heat cooking. Cumin, sesame seed, oregano, sage, thyme, rosemary, and leek are found to be suitable for deep-frying. These spices (except for sesame) are characterized by having a strong flavor, which may be made milder by interaction with heat.

For making sauces that are not heated, mustard, horseradish, Japanese pepper, turmeric, red pepper, caraway, parsley, and celery show high suitability. Many of these spices are selected for their pungent characteristics. Anise, horseradish, tarragon, mustard, Japanese pepper, mace, celery, allspice, dill, nutmeg, coriander, and red pepper are especially suitable for pickling foods. Most of these

The Patterning Theory of Spice Use

TABLE 4.40 Suitability of Spices Used to Deodorize/Mask for Different Cooking Methods

Spice	Heating					Nonheating		
	SI	BK	FR	ST	DF	DR	PK	FS
Garlic	○	⊙	⊙	○	○	○	○	○
Onion	⊙	⊙	⊙	○	○	⊙	○	○
Bay leaves	⊙	⊙	○	○	○	○	○	
Clove	⊙	○	⊙	○	○	○	○	
Nutmeg	○	⊙	⊙	⊙	○		⊙	○
Leek	○	⊙	⊙	○	⊙	○	○	○
Thyme	⊙	○	⊙	⊙	⊙			
Rosemary	⊙	⊙	⊙	○	⊙			
Caraway	○	○	○	○	○	⊙	○	○
Sage	○	⊙	○	⊙	⊙	○		
Oregano	○	⊙	⊙	○	⊙		○	○
Savory	○	○	○	○	○	○	○	○
Coriander	○	○	○	○	○	○	⊙	⊙

Suitability: ○ < ⊙.
Abbreviations as for Table 4.5.

have either a pungent function or a flavoring function. Coriander, whose major function is a deodorizing/masking function, is used in pickled foods to enhance their appearance.

Mint, horseradish, parsley, tarragon, basil, Japanese pepper, star anise, mustard, coriander, celery, fennel, vanilla, and cardamom all show high suitability

TABLE 4.41 Suitability of Pungent Spices for Different Cooking Methods

Spice	Heating					Nonheating		
	SI	BK	FR	ST	DF	DR	PK	FS
Pepper	○	⊙	○	○	○	○	○	○
Red pepper	○	○	○	○	○	⊙	⊙	○
Mustard						⊙	⊙	⊙
Ginger	○	○	⊙	⊙	○	○	○	○
Horseradish						⊙	⊙	⊙
Japanese pepper						⊙	⊙	⊙

Suitability: ○ < ⊙.
Abbreviations as for Table 4.5.

TABLE 4.42 Suitability of Spices Used as Colorants for Different Cooking Methods

	Heating					Nonheating		
Spice	SI	BK	FR	ST	DF	DR	PK	FS
Paprika	○	○	○	○	○	○	○	○
Turmeric	○	○	⊙	⊙	○	⊙	○	
Saffron	⊙							

Suitability: ○ < ⊙.
Abbreviations as for Table 4.5.

TABLE 4.43 Spices Used for Cooking with Heat

Function	Spices
Flavoring	Parsley, cinnamon, allspice, dill, mint, tarragon, cumin, marjoram, star anise, basil, anise, mace, nutmeg, fennel, sesame seed, vanilla, fenugreek, cardamom, celery
Deodorizing/Masking	Garlic, savory, bay leaves, clove, leek, thyme, rosemary, caraway, sage, oregano, onion, coriander
Pungency	Ginger, red pepper, pepper
Coloring	Paprika, turmeric, saffron

TABLE 4.44 Spices Used for Cooking Without Heat

Function	Spices
Flavoring	Parsley, cinnamon, allspice, dill, mint, tarragon, cumin, marjoram, star anise, basil, anise, mace, nutmeg, fennel, vanilla, cardamom, celery
Deodorizing/Masking	Garlic, savory, bay leaves, clove, leek, caraway, sage, oregano, onion, coriander
Pungency	Japanese pepper, mustard, ginger, horseradish, red pepper, pepper
Coloring	Paprika, turmeric

TABLE 4.45 Spices Used for Cooking by Boiling

Function	Spices
Flavoring	Parsley, cinnamon, allspice, dill, mint, tarragon, cumin, marjoram, star anise, basil, mace, nutmeg, fennel, vanilla, fenugreek, cardamom, celery
Deodorizing/Masking	Savory, leek, caraway, sage, oregano, onion, coriander
Pungency	Ginger, red pepper, pepper
Coloring	Paprika, turmeric, saffron

TABLE 4.46 Spices Used for Baking

Function	Spices
Flavoring	Cinnamon, allspice, dill, cumin, marjoram, anise, mace, nutmeg, fennel, sesame seed, vanilla, fenugreek, cardamom, celery
Deodorizing/ Masking	Garlic, savory, bay leaves, clove, leek, thyme, rosemary, caraway, sage, oregano, onion, coriander
Pungency	Ginger, red pepper, pepper
Coloring	Paprika, turmeric

for use in fresh foods. The kinds of spices selected are very similar to those used for pickled foods (except for vanilla, mint, and parsley, which are suitable for fresh foods but not for pickled foods). Spice suitability according to major function is shown in Tables 4.45–4.52. Table 4.53 summarizes spice suitability for fresh foods, including subfunction as well as major function.

F. Spice Suitability by Nation

Suitability profile of 40 spices analyzed by frequency patterning analysis are as follows:

1. *Parsley*: This spice is used primarily in Western cooking. It is less often used in Eastern cooking, except for Chinese and Southeast Asian cuisine, where it tends to be used as an alternative to coriander leaves, which is also called Chinese parsley (Fig. 4.9).
2. *Celery*: This spice is far more suitable for Western cooking than Eastern cooking, except for Southeast Asia. It is used particularly effectively in Italian and French cooking (Fig. 4.10).
3. *Cinnamon*: This spice is suitable for both Western and Eastern dishes, but is used most extensively in Indian and Southeast Asia cooking. Cinnamon's suitability for Italian cooking is especially low (Fig. 4.11).

TABLE 4.47 Spices Used for Frying

Function	Spices
Flavoring	Cinnamon, allspice, dill, cumin, mace, nutmeg, fennel, sesame seed, fenugreek, cardamom, celery
Deodorizing/ Masking	Garlic, savory, bay leaves, clove, leek, thyme, rosemary, caraway, sage, oregano, onion, coriander
Pungency	Ginger, red pepper, pepper
Coloring	Paprika, turmeric

Table 4.48 Spices Used for Steaming

Function	Spices
Flavoring	Cinnamon, allspice, dill, cumin, marjoram, mace, nutmeg, fennel, sesame seed, vanilla, fenugreek, cardamom
Deodorizing/ Masking	Garlic, savory, bay leaves, clove, leek, thyme, rosemary, caraway, sage, oregano, onion, coriander
Pungency	Ginger, red pepper, pepper
Coloring	Paprika, turmeric

Table 4.49 Spices Used for Deep-Frying

Function	Spices
Flavoring	Parsley, cinnamon, allspice, dill, cumin, marjoram, mace, fennel, sesame seed, vanilla, fenugreek, cardamom, celery
Deodorizing/ Masking	Garlic, savory, bay leaves, clove, leek, thyme, rosemary, caraway, sage, oregano, onion, coriander
Pungency	Ginger, red pepper, pepper
Coloring	Paprika, turmeric

Table 4.50 Spices Used for Sauces (no heat)

Function	Spices
Flavoring	Parsley, cinnamon, allspice, dill, mint, tarragon, cumin, marjoram, basil, fennel, vanilla, cardamom, celery
Deodorizing/ Masking	Garlic, savory, bay leaves, clove, leek, caraway, sage, onion, coriander
Pungency	Japanese pepper, mustard, ginger, horseradish, red pepper, pepper
Coloring	Paprika, turmeric

Table 4.51 Spices Used for Pickling

Function	Spices
Flavoring	Parsley, cinnamon, allspice, dill, mint, tarragon, cumin, marjoram, star anise, basil, anise, mace, nutmeg, fennel, vanilla, cardamom, celery
Deodorizing/ Masking	Garlic, savory, bay leaves, clove, leek, caraway, oregano, onion, coriander
Pungency	Japanese pepper, mustard, ginger, horseradish, red pepper, pepper
Coloring	Paprika, turmeric

The Patterning Theory of Spice Use

TABLE 4.52 Spices (major function) Used with Fresh Foods

Function	Spices
Flavoring	Parsley, cinnamon, allspice, dill, mint, tarragon, cumin, marjoram, star anise, basil, anise, mace, nutmeg, fennel, vanilla, cardamom, celery
Deodorizing/ Masking	Garlic, savory, leek, caraway, oregano, onion, coriander
Pungency	Japanese pepper, mustard, ginger, horseradish, red pepper, pepper
Coloring	Paprika

4. *Allspice*: This spice is used mostly in Western cooking and is less suitable for Eastern cooking. High numerical values are observed for British, American, and German cooking (Fig. 4.12).
5. *Dill*: This spice is used primarily in Western cooking (e.g., American). It is used in almost no Eastern nations except India (Fig. 4.13).
6. *Mint*: This spice is most suitable for Western cooking (e.g., British and American). Only Indian among the Eastern cuisines shows suitability for this spice (Fig. 4.14).
7. *Tarragon*: This spice is best suited for Western cooking (e.g., American and French) and shows almost no suitability for Eastern cooking (Fig. 4.15).
8. *Cumin*: This spice is most suitable for Eastern cooking (e.g., Indian and Southeast Asian) and does not show suitability for any Western cooking (except American) (Fig. 4.16).

TABLE 4.53 Spices (including subfunction) Used in Fresh Foods

Function	Spices
Flavoring	Parsley, cinnamon, allspice, dill, mint, tarragon, cumin, marjoram, star anise, basil, anise, mace, nutmeg, fennel, vanilla, cardamom, celery, coriander, mustard, horseradish, garlic, leek, Japanese pepper, caraway, oregano, savory, pepper, ginger
Deodorizing/ Masking	Garlic, savory, leek, caraway, oregano, coriander, parsley, celery, mint, tarragon, star anise, fennel, vanilla, mustard, horseradish, Japanese pepper, cinnamon, allspice, cumin, mace, nutmeg, pepper, ginger
Pungency	Japanese pepper, mustard, ginger, horseradish, red pepper, pepper
Coloring	Paprika, red pepper

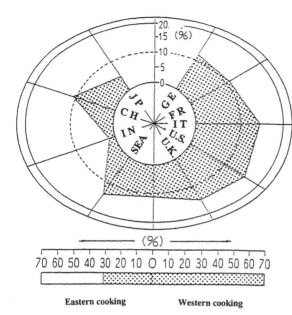

FIGURE 4.9 Suitability pattern for parsley.

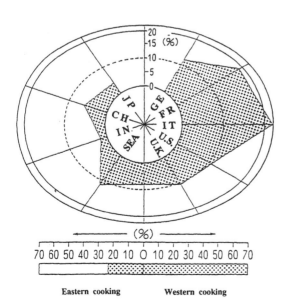

FIGURE 4.10 Suitability pattern for celery.

The Patterning Theory of Spice Use

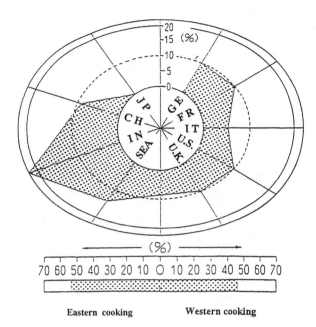

FIGURE 4.11 Suitability pattern for cinnamon.

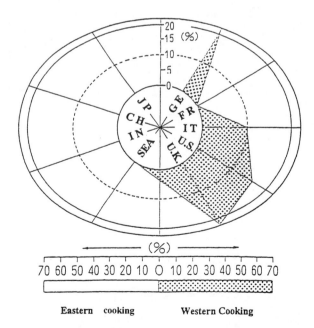

FIGURE 4.12 Suitability pattern for allspice.

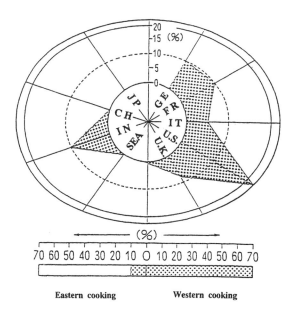

FIGURE 4.13 Suitability pattern for dill.

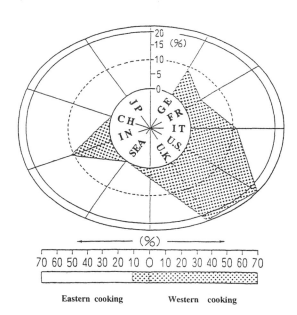

FIGURE 4.14 Suitability pattern for mint.

The Patterning Theory of Spice Use

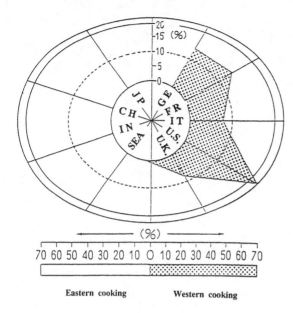

FIGURE 4.15 Suitability pattern for tarragon.

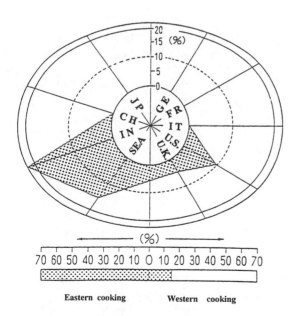

FIGURE 4.16 Suitability pattern for cumin.

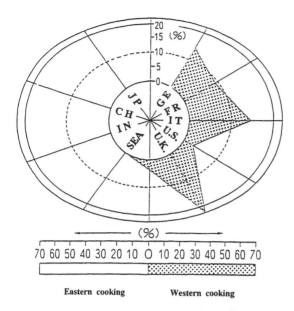

FIGURE 4.17 Suitability pattern for marjoram.

9. *Marjoram*: This spice is most suitable for Western cooking (e.g., British, Italian, and German) and does not show suitability for Eastern cooking (Fig. 4.17).
10. *Star anise*: This spice is most suitable for Eastern cooking (e.g., Chinese and Southeast Asian) and does not show suitability for Western cooking (Fig. 4.18).
11. *Basil*: This spice is most suitable for Western cooking (e.g., American and Italian) and shows almost no suitability for Eastern cooking (Fig. 4.19).
12. *Anise*: This spice is most suitable for Western cooking (e.g., American) and shows almost no suitability for Eastern cooking except for Indian cooking (Fig. 4.20).
13. *Mace*: This spice is most suitable for Indian cooking, and shows almost no suitability for Western cooking except in British foods (Fig. 4.21).
14. *Fennel*: This spice is much more suitable for Indian and British cooking than any other nation's cooking (Fig. 4.22).
15. *Sesame seed*: This spice is most suitable for Eestern cooking (e.g., Japanese and Chinese) (Fig. 4.23).

The Patterning Theory of Spice Use

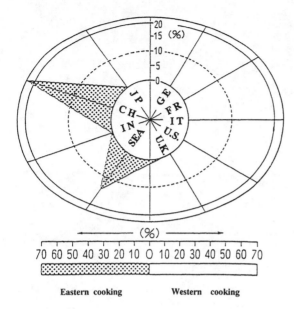

FIGURE 4.18 Suitability pattern for star anise.

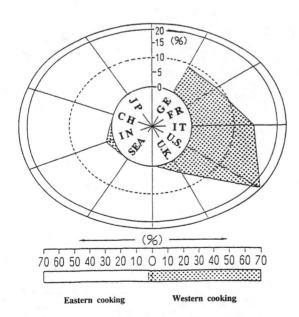

FIGURE 4.19 Suitability pattern for basil.

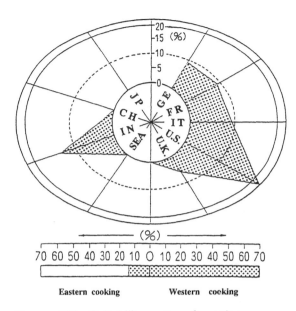

FIGURE 4.20 Suitability pattern for anise.

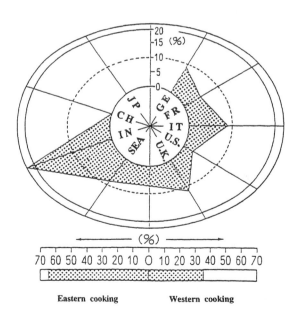

FIGURE 4.21 Suitability pattern for mace.

The Patterning Theory of Spice Use

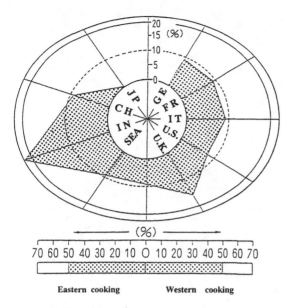

FIGURE 4.22 Suitability pattern for fennel.

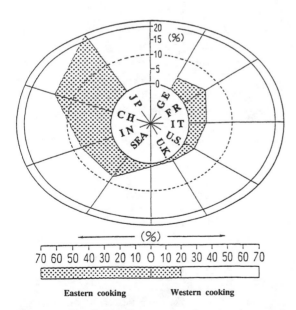

FIGURE 4.23 Suitability pattern for sesame seed.

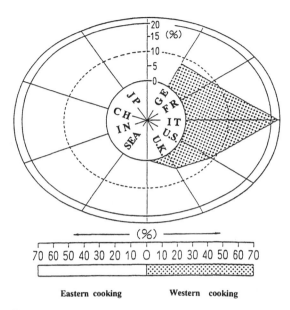

FIGURE 4.24 Suitability pattern for vanilla.

16. *Vanilla*: This spice is used for Western cooking, especially Italian cooking. It shows almost no suitability for Eastern cooking (Fig. 4.24).
17. *Fenugreek*: This spice does not show great suitability for any except for Indian cooking (Fig. 4.25).
18. *Cardamom*: This spice is most suitable for Indian cooking, but does not show above-average suitability for any other national cuisine (Fig. 4.26).
19. *Garlic*: This spice is almost equally suited to Western and Eastern cooking. Individually, it shows slightly higher suitability for Italian and Chinese cooking (Fig. 4.27).
20. *Onion*: This spice is found to be suitable for both Eastern and Western cooking, especially American and French cooking (Fig. 4.28).
21. *Bay leaf*: This spice is most suitable for Western cooking (e.g., French and Italian) and does not show great suitability for any Eastern cuisine except Southeast Asian cooking (Fig. 4.29).
22. *Clove*: This spice is overwhelmingly suited to Indian cooking (Fig. 4.30).
23. *Nutmeg*: This spice is far more suitable for Western cooking (e.g., German and British) than Eastern cooking except for Indian (Fig. 4.31).

The Patterning Theory of Spice Use

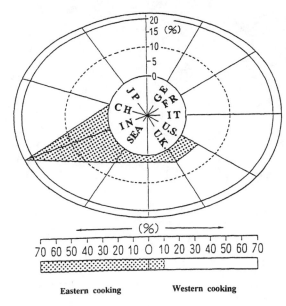

FIGURE 4.25 Suitability pattern for fenugreek.

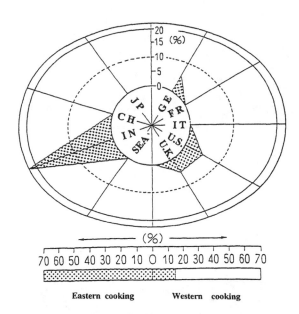

FIGURE 4.26 Suitability pattern for cardamom.

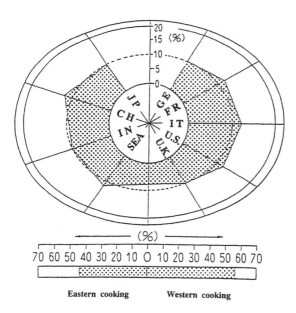

FIGURE 4.27 Suitability pattern for garlic.

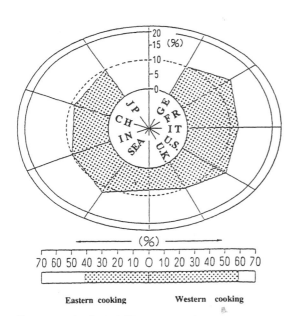

FIGURE 4.28 Suitability pattern for onion.

The Patterning Theory of Spice Use

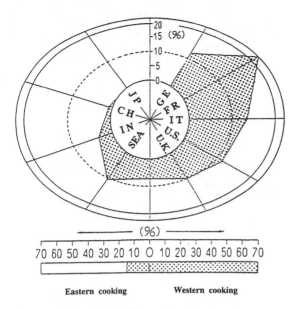

FIGURE 4.29 Suitability pattern for bay leaves.

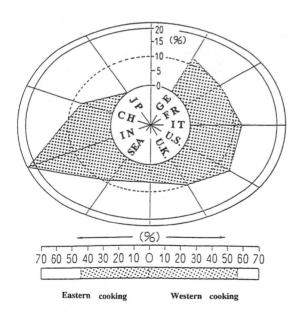

FIGURE 4.30 Suitability pattern for clove.

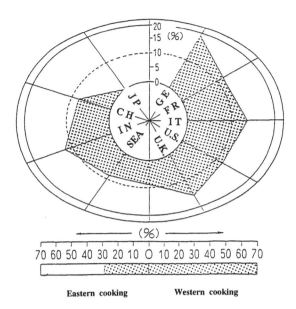

FIGURE 4.31 Suitability pattern for nutmeg.

24. *Leek*: This spice is more suitable for Italian, Japanese, and Southeast Asian cooking than for other national cuisines (Fig. 4.32).
25. *Thyme*: This spice is far more suitable for Western cooking, especially French and Italian, than for Eastern cooking (Fig. 4.33).
26. *Rosemary*: This spice is far more suitable for Western cooking, especially French, British, and Italian, than for any Eastern cooking (Fig. 4.34).
27. *Caraway*: This spice is suitable for Western cooking, especially German (Fig. 4.35).
28. *Sage*: This spice is suitable for Western cooking, especially British (Fig. 4.36).
29. *Oregano*: This spice is suitable for Western cooking, especially Italian (Fig. 4.37).
30. *Savory*: This spice is much used for Western cooking, especially German (Fig. 4.38).
31. *Coriander*: This spice is most suitable for Eastern cooking, especially Chinese and Indian (Fig. 4.39).
32. *Pepper*: There is almost no difference in suitability of this spice between Eastern and Western cooking, but it shows very high suitability for American cooking (Fig. 4.40).

The Patterning Theory of Spice Use

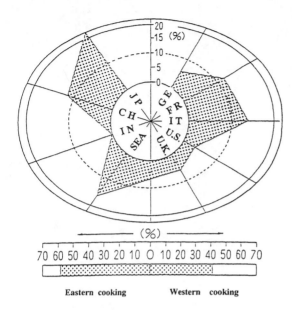

FIGURE 4.32 Suitability pattern for leek.

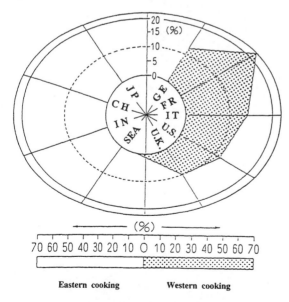

FIGURE 4.33 Suitability pattern for thyme.

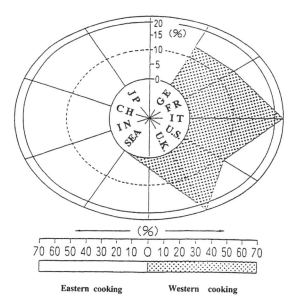

FIGURE 4.34 Suitability pattern for rosemary.

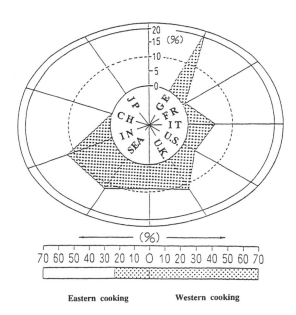

FIGURE 4.35 Suitability pattern for caraway.

The Patterning Theory of Spice Use

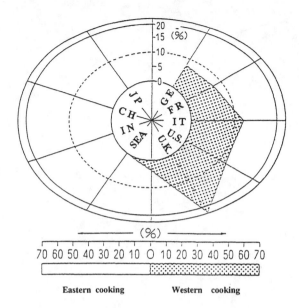

FIGURE 4.36 Suitability pattern for sage.

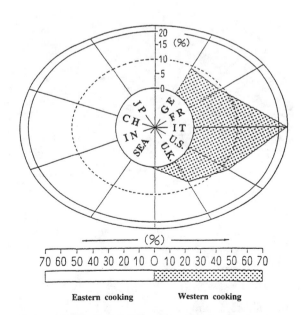

FIGURE 4.37 Suitability pattern for oregano.

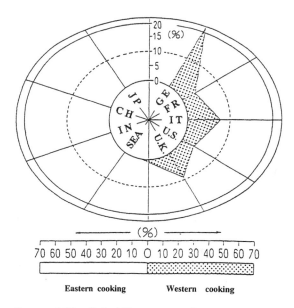

FIGURE 4.38 Suitability pattern for savory.

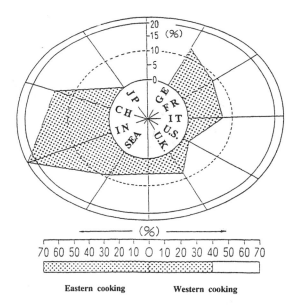

FIGURE 4.39 Suitability pattern for coriander.

The Patterning Theory of Spice Use

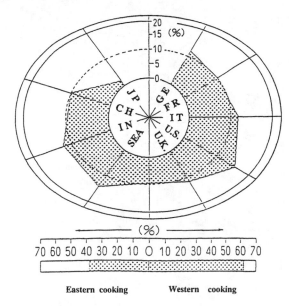

FIGURE 4.40 Suitability pattern for pepper.

33. *Red pepper*: This spice is suitable for Eastern cooking, especially for Southeast Asian and Indian (Fig. 4.41).
34. *Mustard*: This spice is suitable for both Western and Eastern cooking, and shows a slightly higher suitability for Japanese cooking (Fig. 4.42).
35. *Ginger*: This spice is most suitable for Eastern cooking, especially Chinese and Japanese (Fig. 4.43).
36. *Horseradish*: This spice is suitable for Japanese cooking among Eastern nations and for British cooking among Western ones (Fig. 4.44).
37. *Japanese pepper*: This spice is suitable for Eastern cooking, especially Japanese (Fig. 4.45).
38. *Paprika*: This spice is suitable for Western cooking (e.g., American and German), but it does not show suitability for Eastern cooking except for Southeast Asian (Fig. 4.46).
39. *Turmeric*: This spice is suitable for Eastern cooking, especially Indian (Fig. 4.47).

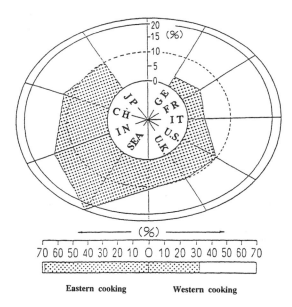

FIGURE 4.41 Suitability pattern for red pepper.

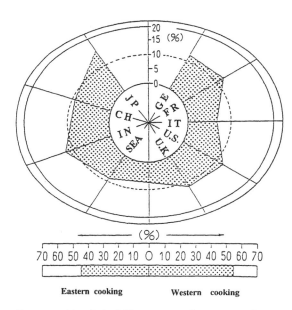

FIGURE 4.42 Suitability pattern for mustard.

The Patterning Theory of Spice Use

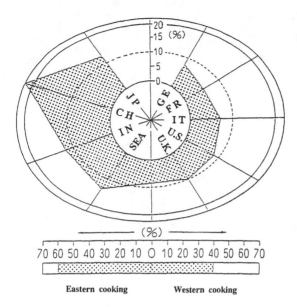

FIGURE 4.43 Suitability pattern for ginger.

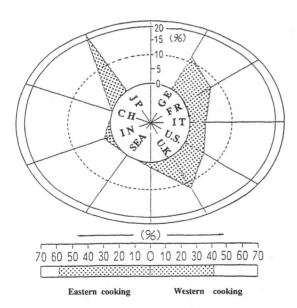

FIGURE 4.44 Suitability pattern for horseradish.

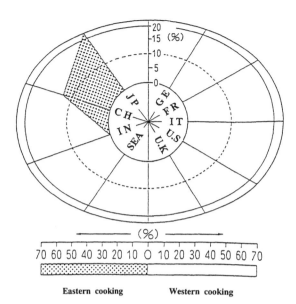

FIGURE 4.45 Suitability pattern for Japanese pepper.

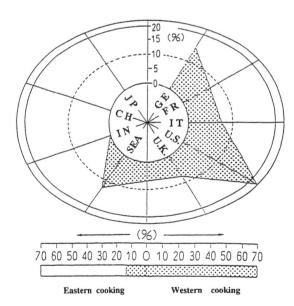

FIGURE 4.46 Suitability pattern for paprika.

The Patterning Theory of Spice Use

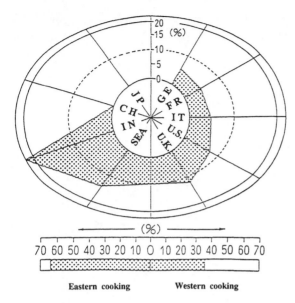

FIGURE 4.47 Suitability pattern for turmeric.

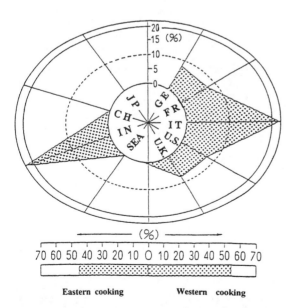

FIGURE 4.48 Suitability pattern for saffron.

40. *Saffron*: This spice is suitable for Indian and Italian cooking and does not show suitability for any other nation's cuisine (Fig. 4.48).

REFERENCE

1. M. Takamasa, *Nippon Koshinryo Kenkyuukai Koen Yoshisyu*, (8): (1993).

5
Physiological Effects of Spice Components

I. FUNDAMENTAL PHYSIOLOGICAL EFFECTS OF PUNGENT SPICES

It is well known that the human body is physiologically affected by the ingestion of pungent spices, and several studies have been conducted on the nutritive biochemical effects of the intake of spice components.

A. Absorption of Pungent Compounds

Kim et al. observed how much capsaicin and its analogs contained in red pepper were absorbed when taken in as components of food and found that only a small quantity of these compounds were absorbed in the gut [1]. In contrast, Kawada et al. reported that capsaicin and dihydrocapsaicin were absorbed in the gastrointestinal (GI) tract of rats within a few hours [2]. In his study, male Wistar rats were administered 3 mg capsaicin by stomach tube. The absorption of capsaicin in the GI tract was confirmed by measuring the amount of residue in the stomach, duodenum, jejunum, ileum, cecum, and large intestine. It was found that both capsaicin and dihydrocapsaicin were absorbed in the stomach quickly and that both were also absorbed in the small intestine (duodenum and jejunum) (Fig. 5.1). It was also found that over 80% of capsaicin prescribed was absorbed within 3 hours after prescribing. The route of absorption was also confirmed using [^{3}H]-dihydrocapsaicin. Dihydrocapsaicin was absorbed from jejunal to mesenteric venous blood, and it was transported throughout the body by associating with the albumin of serum proteins in mesenteric venous blood. The rates of absorption of other pungent principles were also investigated [3]. Emulsions composed of piperine, the major pungent principle of black pepper, and zingerone, the major pungent principle of ginger, were injected into ligated jejunal loops and their rates of absorption from the loops observed. The amounts of piperine and zingerone absorbed from the loops were 85 and 95%, respectively, in one hour.

B. Heat Production (Thermogenesis) Induced by Pungent Compounds

It has been confirmed that the pungent components of red pepper promote lipid metabolism and induce body heat production [4]. Absorbed capsaicin is believed to result in a release of a neurotransmitter, substance P, from the spinal cord by activation of the primary afferent neurons, which activates sympathetic efferent nerves and enhances the secretion of adrenal catecholamine, especially epinephrine. As shown in Figure 5.2, a significant increase in epinephrine secretion was observed in rats infused with capsaicin [5]. When rats are injected with capsaicin, the level of serum glucose increases rapidly and that of serum free fatty acids increases gradually, while the liver glycogen content decreases (Fig. 5.3) [6]. The reason for this is that catecholamine secreted into the adrenal vein triggers β-adrenergic receptors in the liver and the adipose tissue, which are the major internal organs. This promotes the decomposition of glycogen to glucose and the formation of free fatty acids. As explained later, these compounds are carried by the blood into the terminal tissues, such as muscle, to be burned and, therefore, enhance body heat production (Fig. 5.4). According to Kawada et al., piperine and zingerone also evoked catecholamine secretion from the adrenal medulla, although to a lesser extent than that caused by capsaicin [3]. This means that both piperine and zingerone also have the ability to enhance body heat production. Other volatile pungent compounds, including allylisothiocyanate found in mustard and diallyldisulfide found in garlic, were not found to enhance the secretion of catecholamine (Fig. 5.5).

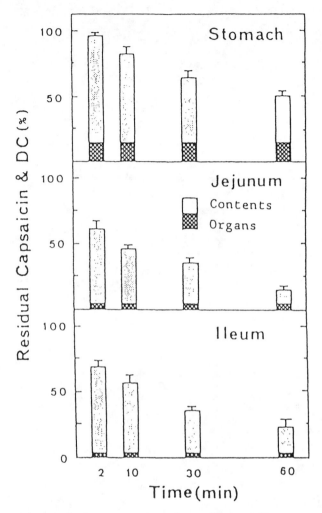

FIGURE 5.1 Disappearance of capsaicin and dihydrocapsaicin from the ligated loops of stomach, jejunum, and ileum in situ. Stomach loops were infused with 1 mM capsaicin mixture, 0.3% Tween 20 in 0.9% saline (1 ml), and jejunum (10 cm) or ileum (10 cm) was incubated with 1 mM capsaicin mixture, 5 mM sodium taurocholate, 5 mM monoolein in phosphate buffer (pH 6.3). (From Ref. 2.)

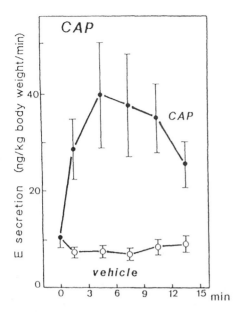

FIGURE 5.2 Time course of epinephrine secretion from the rat adrenal medulla evoked by capsaicin. The rats were infused with capsaicin (200 μg/kg) for 1 minute. (From Ref. 5.)

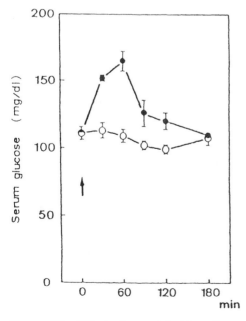

FIGURE 5.3 Effect of capsaicin (4 mg/kg, ●–●) and the vehicle (○–○) on serum glucose level. (From Ref. 6.)

Physiological Effects of Spice Components

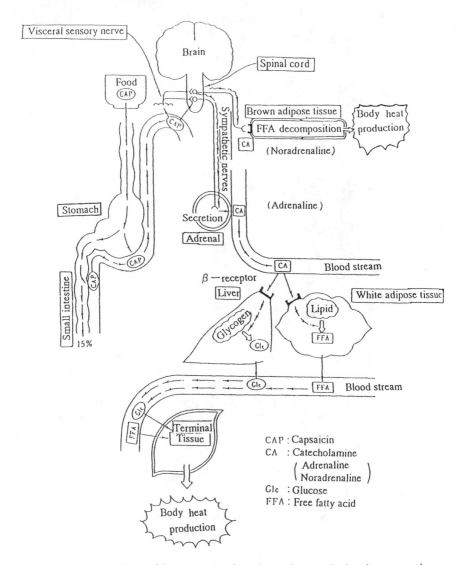

FIGURE 5.4 Induction of heat production through catecholamine secretion.

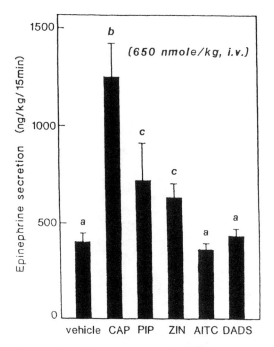

FIGURE 5.5 Effect of capsaicin (CAP), piperine (PIP), zingerone (ZIZ), allylisothiocyanate (AITS), and diallyldisulfide (DADS) on adrenal epinephrine secretion. Rats were intravenously infused with each pungent principle solution (650 nmol/kg) or the vehicle (a 0.9% NaCl solution containing 0.1% ethanol and 0.5% Tween 80) for 1 minute. (From Ref. 3.)

C. Effects of Pungent Spices on Digestion and Absorption

Some pungent spice components are known to have an effect on digestive tract function. From a physiological point of view, these effects could be classified as (1) promotion of chemical digestion by promoting secretion of digestive juices, (2) promotion of physical digestion by promoting digestive tract movement, and (3) promotion of speed of absorption by increasing blood circulation in the digestive tract.

It is known that saliva secretion is increased by the intake of pungent spice. The amount of saliva normally secreted per day is 1–1.5 liters. Saliva is composed of water (>99%), protein (including α-amylase and mucin), electrolytes, and a very small quantity of immunoglobulin. The secretion of saliva was reported to be promoted especially by the nonvolatile compounds of pungent spice [9].

The major movement in the esophagus is peristalsis. This movement is enhanced by the excitation of the vagus nerve caused by the contraction of muscle in the esophagus and is inhibited by the excitation of the sympathetic nerve. Food transferred from the esophagus is mixed with the acidic gastric juice in the stomach. In this process, peristalsis controlled by the vagus nerve also plays an important role in promoting contraction and tonus of the stomach, while the sympathetic nerve cause the inverse. Some spice components could have an effect on the movement of the digestive tract. Capsaicin, in particular, affects the nerves of the small intestine, releasing substance P. This neurotransmitter works directly on the muscle of intestine to affect the cholinergic nerves. It is believed that acetylcholine is secreted as a result, affecting the vagus nerve. Besides capsaicin, piperine and zingerone both affect the vagus nerve.

In the blood stream, nonvolatile pungent compounds including capsaicin promote the secretion of norepinephrine, which causes contraction of blood vessels. This results in an increase in blood circulation in the digestive tract.

D. Effect of Spices on the Circulatory System

Blood pressure is controlled by pulsation of the heart, resistance to blood circulation, and elasticity of the arterial vessel. Smooth muscle in the blood vessel walls is controlled by the vasoconstrictor nerve, a sympathetic nerve. Vasoconstriction causes secretion of noradrenaline (norepinephrine), which in turn increases the resistance of the blood vessels. On the other hand, the sympathetic nerve, which controls the action of the heart, increases the heart rate and strengthens the contractile force of the heart and the amount of blood pumped out. As a result, a large amount of blood flows into a contracted blood vessel, resulting in raised blood pressure. The vagus nerve works conversely, that is, its stimulation decreases blood pressure. Spice compounds do not affect the sympathetic nerve of the heart if ingested in ordinary amounts, but such amounts will increase or decrease the blood pressure of peripheral blood vessels, resulting in a change in the amount of blood flow. Norepinephrine secretion was also observed when capsaicin, piperine, and zingerone were infused, although the level of secretion was relatively low, compared with the amount of epinephrine secreted [3]. It is believed that the rise in the temperature of the body's surface caused by the intake of some spices is due to transfer of body heat from inside the body to the body surface caused by the increase in the amount of blood flow.

E. Effect of Spices on Metabolic Regulation

Metabolism can be divided into anabolism, in which complex material is synthesized from simple substances, and catabolism, in which the reverse occurs. In addition to the macronutrients, such as proteins, fats, and carbohydrates, certain minor constituents such as those in coffee and tea are known to have an effect on metabolic rate. If the process of digestion and absorption is thought of as anabolism in the broad sense, many spice compounds are thought to promote this process. Pungent components are also involved in catabolism. It has become clear that pungent spices increase the activity of the sympathetic nerve, which results in the activation of the following two catabolic channels:

1. sympathetic nerve → adrenal gland → secretion of adrenaline
2. sympathetic nerve → brown adipose tissue → secretion of noradrenaline

This process would promote catabolic events, including lipid metabolism, energy metabolism, and generation of body heat. As mentioned, pungent compounds have an effect on metabolic regulation as well as on absorption and digestion. The mechanism of spice-induced thermogenesis is characterized by acting on the metabolism of a living body through an autonomic ganglion and endocrine system, in spite of the fact that these pungent components are not nutrients (Fig. 5.6).

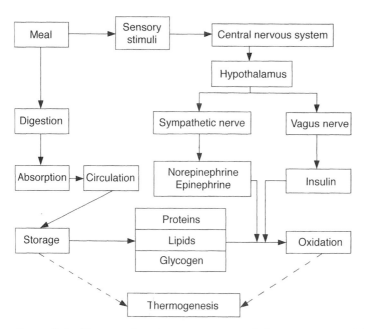

FIGURE 5.6 Hormonal and nervous system influences on meal-induced thermogenesis. Broken lines indicate sources of heat production. (From Ref. 8.)

II. TEMPERATURE CHANGE CAUSED BY INTAKE OF PUNGENT SPICES

The intake of pungent spices not only stimulates the stomach mucosa, it also accelerates peristalsis in the intestine, so that nutrient absorption is increased. An overall dietary improvement would be expected, summarized in the following steps:

1. Pungent spice stimulate mucosa of digestive organ to activate the central nervous system
2. The amount of blood carried to digestive organs is increased
3. Secretion of digestive juices is increased

The temperature of the skin surface increases immediately after ingestion of the pungent spice, although actual body temperature (measured rectally) does not change greatly.

There are differences in how the temperature of the skin surface increases or decreases, depending upon the type of pungent spice and how it is ingested. After a person has eaten a pungent food, the temperature of the skin surface increases gradually. The person will feel hot and may start to perspire, after which the temperature of the skin surface will eventually decrease. If the pungent compound is ingested in large quantity, the temperature of the skin surface will eventually drop lower after sweating than it was before the food was eaten. In general, the more pungent the food, the bigger the difference between the highest temperature recorded before sweating and the lowest temperature after sweating. This phenomenon is credited with the "cooling" effect achieved by the ingestion of pungent foods. See Figure 5.7 for a graphic illustration of this phenomenon.

The body heat of people in tropical and subtropical regions is said not to be easily dissipated because of the hot and humid climate. It is for this reason that in the cuisines of these regions red pepper is utilized in a wide variety of foods—to let the accumulated heat discharge from the skin surface. Pungent spices can be also used in the foods in cold regions to raise body temperature, but it is necessary to control the amount of pungent spice used so that the skin temperature will not drop too greatly afterward.

III. WEIGHT CONTROL EFFECTS OF PUNGENT SPICES

Recently, several studies reported on the dietary effects of pungent spices. It is thought that appropriate intake of pungent spices may aid in weight control, although one might assume that the spice could also increase one's appetite. When a small amount of pungent spice is ingested, energy consumption is increased due

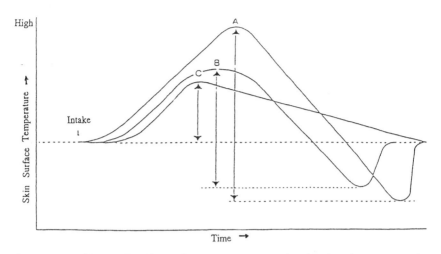

FIGURE 5.7 Change in skin surface temperature after intake of pungent spice. A: High pungency; B: medium pungency; C: low pungency.

to thermogenesis. Actually, when capsaicin, which is the major component of red pepper, is mixed into feed given to rats, their weight tends to decrease even though they eat much more than they do of the "normal feed" containing no capsaicin.

Kawada et al. [10] assumed that the induction of thermogenesis by pungent spice would influence energy metabolism resulting in a decline in the accumulation of somatic lipid. In their study, high-lipid feed composed mainly of lard to which was added 0.014% capsaicin (average daily intake in Southeast Asia) was given to rats for 10 days. They observed the effect of capsaicin on lipid metabolism by comparing these rats with rats fed a noncapsaicin diet. It was found that the weight of perirenal adipose tissue and the serum triglyceride concentration were significantly lower in the rats fed capsaicin than in those not fed capsaicin, and that perirenal lipoprotein lipase activities were increased by supplementing capsaicin in the diet, although the rate of weight gain and the rate of lipid absorption were not affected by capsaicin supplementation (Table 5.1). The weight of perirenal adipose tissue and serum triglyceride tended to decrease with an increase in the amount of capsaicin added (Figs. 5.8 and 5.9). It follows from these results that capsaicin decreases the weight of perirenal adipose tissue and serum triglyceride by stimulating lipid metabolism from adipose tissue.

Henry and Emery [11] studied the effect of spices on metabolism. Twelve panels of 20-year-olds who were accustomed to eating spiced food took part in this study, and metabolic rate was expressed by oxygen consumption. After measuring resting metabolic rate (RMR), half of the panels were given a breakfast that did not contain any spice, and the rest were given the same breakfast to which chili sauce

TABLE 5.1 Effect of Capsaicin Administration on Various Enzyme Activities in Rat Liver and Perirenal Adipose Tissue

	Nonpurified	High fat	High fat plus 0.014% capsaicin	p-value, high fat vs. high fat plus capsaicin
Liver				
Acetyl-CoA carboxylase (mU/g liver)	199.0 ± 14.9	74.7 ± 5.0	77.4 ± 10.2	NS
Glucose-6-phosphate dehydrogenase (U/g liver)	1.04 ± 0.63	0.76 ± 0.1	1.09 ± 0.05	< 0.02
β-Hydroxyacyl-CoA dehydrogenase (mU/g liver)	39.7 ± 2.9	53.0 ± 1.8	53.5 ± 2.9	NS
Perirenal adipose tissue				
Hormone-sensitive lipase (free fatty acid; μmol/μg DNA per h)	3.22 ± 0.25	4.92 ± 0.17	4.45 ± 0.16	NS
Lipoprotein lipase (free fatty acid; μmol/μg DNA per h)	23.0 ± 1.4	20.5 ± 0.3	25.5 ± 0.7	< 0.01

Source: Ref. 10

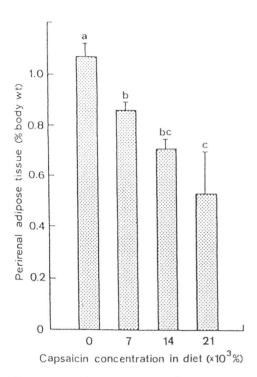

FIGURE 5.8 Dose response of capsaicin on perirenal adipose tissue weight. (From Ref. 10.)

and mustard sauce were added on consecutive days. After each breakfast was ingested, RMR was measured intermittently for 3 hours. The overall average increase in the RMR for 3 hours after ingesting the breakfast with added spice was 158%; that for the other group was 128%. In short, the metabolic rate was increased 25% by adding spice (Fig. 5.10).

Brown adipose tissue (BAT) has an effect on diet-induced thermogenesis, and it is known that functional abnormalities of BAT often cause obesity in animals. Yoshida et al. [12] conducted research to clarify whether pungent principles of spices would activate BAT thermogenesis by measuring interscapular brown adipose tissue (IBAT) temperature and mitochondrial oxygen consumption in BAT in rats into which capsaicin, isothiocyanate, and other pungent compounds were injected intramuscularly. Ephedrine, a compound known to activate BAT function, was also injected for comparison. They found that both capsaicin and isothiocyanate increased IBAT temperature at certain doses (Figs. 5.11 and 5.12) and that oxygen consumption in IBAT increased for rats fed capsaicin and isothio-

Physiological Effects of Spice Components

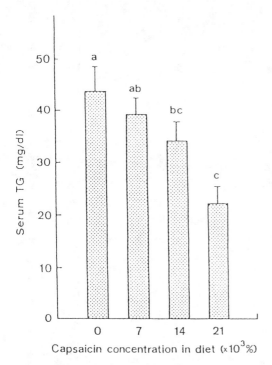

FIGURE 5.9 Dose response of capsaicin on serum triglyceride concentration. (From Ref. 10.)

cyanate as it did in the case of ephedrine, compared with control rats, indicating that both pungent compounds activated BAT function and could be used to mitigate obesity.

Capsaicin and its homolog are absorbed easily in the body and dissolve in blood immediately. As explained, these compounds affect the adrenal gland through the central nervous system to promote the secretion of epinephrine. Glycogen from the liver is broken down and the level of sugar in the blood rises. This sugar serves as an energy source, after which lipid is consumed rather than sugar. When energy metabolism shifts from sugar to lipid, both the level of sugar in the blood and the amount of free fatty acids are high, and the secretion of insulin is depressed due to the secretion of adrenaline. This metabolic process is ideal for endurance sports.

Kim et al. [13] studied the effect of capsaicin on endurance in the mouse. Endurance was evaluated when measuring lipid metabolism was enhanced after feeding capsaicin [13]. In this study, the effect of capsaicin (6 mg/kg) on swimming mice was observed. A mouse that started to swim 3 hours after ingesting

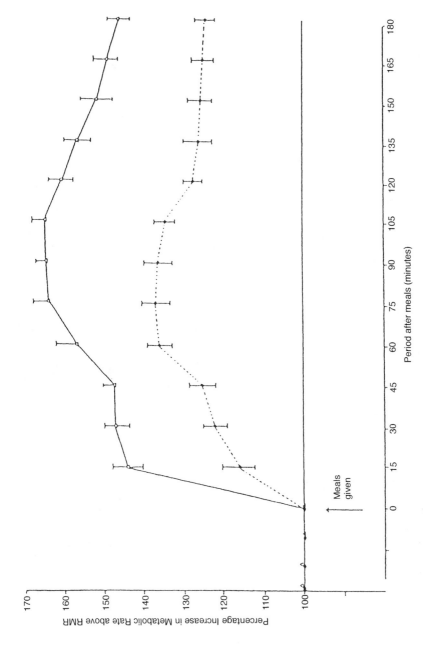

FIGURE 5.10 Mean increase in metabolic rate in subjects fed meal alone (●–●) or meal plus spices (△–△). (From Ref. 11.)

capsaicin was observed to swim significantly longer than when it was not fed capsaicin, but there was almost no effect of capsaicin on mice starting to swim 1 or 3 hours after feeding.

It can be concluded that red pepper or its pungent principles could have a significant effect on preventing animals (and humans) from becoming overweight not only by enhancing lipid metabolic rate and lowering the weight of perirenal adipose tissue, but also by strengthening endurance, which could increase activity. It is also possible that capsaicin and isothiocyanate could mitigate obesity, although the effect would need to be confirmed by long-term administration of these compounds. As mentioned above, some other pungent compounds including piperine and zingerone were also found to promote secretion of epinephrine and could be expected to inhibit fat formation in the body.

IV. PHYSIOLOGICAL ACTIVITY OF LILIACEAE FAMILY SPICES

A. Prevention of Thrombus Formation

Allicin and diallylsulfide are sulfide compounds converted from alliine existing in garlic tissue. Intake of these compounds increases the appetite, suppresses the growth of microbes in the intestine, and promotes blood circulation. Allicin is characterized by being easily absorbed from the digestive organs and by staying in the digestive organs for a long period of time. Allicin is also known to suppress aneurinase, an enzyme that decomposes vitamin B_1.

Around 16 compounds have been identified in garlic, including alliine, allicin, and scordinin. These compounds degrade or decompose upon heating to generate several compounds, including ajoene, which, when it coexists with methyl allyltrisulfide (MATS), acts to hinder thrombus production by preventing platelet aggregation. Also, scordinin was confirmed to decrease cholesterol levels in rabbits [14]. It was also confirmed that scordinin lowers both glutamic oxaloacetic transaminase (GOT) and glutamic pyruvic transaminase (GPT) levels in blood. This would indicate that scordinin has an ability to maintain normal liver function.

Blood aggregation functions to protect the body by preventing blood loss when there is an external injury. This phenomenon can also occur inside the body. In case of an injury to microvessels in the brain or blood vessels of the heart, platelets in the blood aggregate as they would around an open wound. A blockade or stricture tends to occur when high amounts of cholesterol or saturated fatty acids are ingested. Once blood platelets aggregate, the blood starts to coagulate around the core of the blood platelet aggregation, and this aggregated blood is then carried to the site of the stricture. The flow of blood from the thrombus would then slow and eventually stop, resulting in myocardial or brain infarction.

Ajoene compounds contained in onion and garlic have been attracting

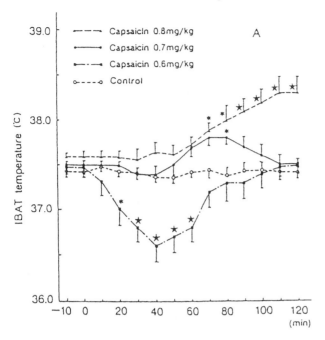

FIGURE 5.11 Effect of intramuscular capsaicin injection on (A) IBAT and (B) rectal temperature. (From Ref. 12.)

attention for their prevention of thrombus formation. There are other reports about the compounds preventing thrombus formation [15,16]. A study on various compounds in garlic found that MATS prevented platelet aggregation. Platelets aggregate when arachidonic acid is metabolized to produce thromboxane A_2. But MATS in garlic inhibits thromboxane formation to prevent platelet aggregation (Fig. 5.13). MATS is included in the essential oil of garlic by only 5%. Oral intake of garlic or onion is said to enhance the ability to separate and dissolve blood thrombi and to be effective in the treatment of ischemic heart disease.

Recently, eicosapentaenoic acid (EPA) has begun to attract attention for its ability to inhibit the first step of blood aggregation. Saffron (which does not belong to the Liliaceae family) was reported to inhibit platelet aggregation [17]; its active substance has been identified as adenosine.

B. Other Effects of Garlic

Garlic has been known to increase sexual desire [14]. This is considered to be due to (1) the immediate effect of absorbed allicin to stimulate the mucosa of the genitals when it is discharged from the urethra, and to (2) the delayed effect that

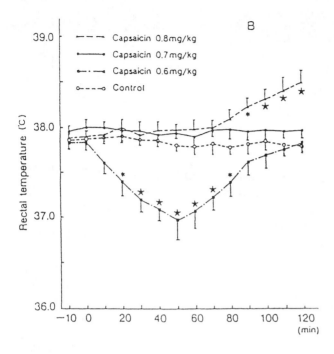

scordinin absorbed slowly into the body activates cells throughout the body. Formation of spermatozoa in a spermary of rats to which scordinin was fed was more apparent than in rats not fed scordinin. It was also confirmed that very large amounts of spermatozoa were produced in rats fed scordinin-containing feed. The effect of increased sexual desire by scordinin is therefore thought to be due to activation of the male organs. It was also found that the swimming time of rats was extended greatly when scordinin was included in their feed.

Another research group found the death rate for rats fed scordinins was also found to be much lower than that for rats not fed scordinin when both groups were kept in high-temperature conditions. From these studies, it is estimated that scordinin increases stamina and improves resistance to stress.

C. Effects of Sulfur-Containing Spice Compounds

As mentioned in a previous chapter, many pungent and flavor compounds found in members of the *Allium* genus, such as garlic and onion, are sulfur-containing compounds. Sulfur compounds function to (1) lower the level of sugar in blood, (2) lower the level of cholesterol in blood, (3) dissolve thrombi, (4) hinder platelet

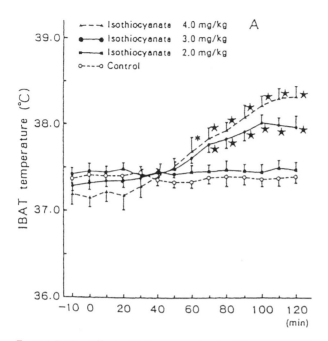

FIGURE 5.12 Effect of intramuscular isothiocyanate injection on (A) IBAT and (B) rectal temperature. (From Ref. 12.)

aggregation, (5) prevent hyperlipemia, (6) prevent arthritis, and (7) promote diuresis.

Garlic also has the ability to detoxify hydrocyanic acid. Tazoe et al. studied this effect in mice [18]. They injected garlic extract intravenously, administered potassium cyanide (150 µg/kg body weight) to the mice, and recorded the death rate. More mice given the garlic extract were found to remain alive after a certain period of time than mice receiving the potassium cyanide only. The detoxification mechanism results from the formation of rhodan during the reaction of hydrogen cyanide with the sulfur in garlic. The amount of rhodan produced after giving either allicin or garlic juice for 2 weeks was observed to enhance this mechanism.

V. ANTI-HIV EFFECTS OF LABIATAE FAMILY SPICES

Acquired immunodeficiency syndrome (AIDS), caused by the human immunodeficiency virus (HIV), is currently treated with a number of drugs and drug combinations, some of which have unpleasant or harmful side effects. Yamazaki

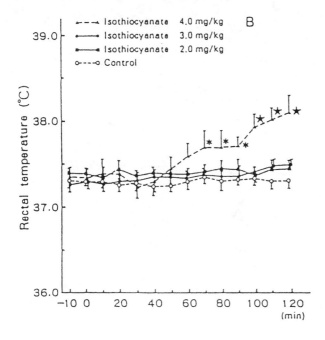

[19] studied the effect of both medicinal and edible herbs and plants of the Labiatae family against HIV. In this study, ground herbs or Labiatae plants were extracted by acetone, ethanol, and 70% ethanol, in that order, and their residues were freeze-dried (water extract). Each spice extract was added at various concentrations to MT-4 cells infected partially by HIV (LAV-1) and was cultivated for 5 days. This study found no effects for acetone and ethanol extracts containing flavor compounds. But suppressive effects were found for water extracts of 30 plants and 70% ethanol extracts of some plants (see Table 5.2). Most of these spices showed effects at relatively low concentrations (<31 μg/ml). He also tested the effects on both LAV-1 and KK-1 virus obtained directly from an AIDS patient and found that lemon balm in particular was effective for both viruses.

Yamazaki [19] also studied whether these effective extracts inhibited infection from cell to cell using giant cells for the purpose of clarifying the mechanism of the extracts' anti-HIV function. The principle of this experiment was that a giant cell would be formed if noninfected cells clung to the outside of the HIV-infected cell. They observed microscopically whether giant cells were formed after adding each plant extract to the mixture of noninfected Molt-4 cells (known

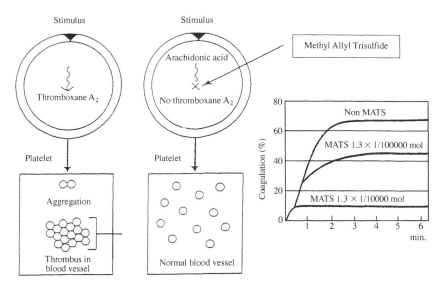

Figure 5.13 Suppression of platelet aggregation.

Table 5.2 Anti-HIV Function of Different Spice Extracts

Spice	Water extract IC_{100} (*1)	70% ethanol extract IC_{100}
Peppermint	62	125
Thyme	62	N
Rosemary	125	N
Marjoram	31	N
Lavender	31	500
Oregano	31	N
Spearmint	31	N
Basil	16	1000
Lemon balm	16	N
Basil	31	125
Perilla	31	N

IC_{100}(*1): 100% inhibitory concentration for HIV-1 infection on MT-4 cell (μg/ml).
N: No inhibitory action for HIV-1.
Source: Adapted from Ref. 19.

TABLE 5.3 Inhibitory Concentration of Water Extracts of Different Spices

Water extract of spice	IC_{100} for kk-1/MT-4 cell (μg/ml)	IC_{100} for giant cell (μg/ml)
Marjoram	—	125
Lavender	250	62.5
Oregano	—	62.5
Spearmint	—	125
Basil-cinnamon	125	62.5
Lemon balm	62.5	62.5
Basil-cinnamon	250	125
Perilla	125	62.5

Source: Adapted from Ref. 19.

to form a giant cell easily) and HIV-infected Molt-4 cells. As Table 5.3 shows, formation of giant cells was found to be inhibited by each concentration of extract. These extracts are considered to block HIV virus from being absorbed by the cell.

It follows that many of the plants belonging to the Labiatae family have inhibitory effects on HIV propagation (especially basil-cinnamon, lemon balm, and perilla) and that the effective substances seem to be contained in their water extract, not in their essential oils. It can be surmised that the mechanism of their anti-HIV activity is due mainly to their inhibitory action on cell-to-cell adsorption. It is also possible that any side effects of these water extracts will be less harmful than those experienced with, say, AZT. In other words, it can be estimated that the long-term safety is high, and their future use can be expected.

VI. PHYSIOLOGICAL EFFECTS OF OTHER SPICES

Nutmeg is an evergreen tree belonging to the Myristica family, whose seeds are used as a spice. The skin surrounding the seed is utilized as a different spice—mace. The major compounds in the essential oil of nutmeg are pinene and camphene, accounting for 80% of this oil, and other flavor compounds include geraniol, eugenol, dipentene, and linalool. The essential oil of nutmeg has both antimicrobial and antioxidant activities, as explained in a later chapter, and it has been also reported that a compound found in mace has an antitumor effect [21]. This substance is the phenol compound MSP, which was obtained by extracting mace with acetone followed by fractionating with another solvent. An inhibitory effect on the propagation of cancer cells in vitro was observed in mice, as was an anticancer effect. Research is continuing on the effect of MSP on different kinds of cancers and on its safety and its toxicity, and it is expected that clinical trials will be conducted in a near future. A chemical synthesis of MSP is also considered to be possible, judging from its structure. It is expected that this substance will be widely used in the future if its effect and safety can be confirmed.

In addition to mace, basil, sage, oregano, peppermint, and perilla are believed to have anticancer effects [20]. In this study mice injected with breast cancer–inducing cells were given feed with and without the above spices. Mice ingesting the spice compounds were found to have extended life spans.

REFERENCES

1. N. D. Kim and C. Y. Park, *J. Pharm. Soc. Korea*, 25: 101 (1981).
2. T. Kawada, T. Suzuki, M. Takahashi, and K. Iwai, *Toxicol. Appl. Pharmacol.*, 72: 449 (1984).
3. T. Kawada, S. Sakabe, T. Watanabe, M. Yamamoto, and K. Iwai, *Proc. Soc. Exp. Biol. Med.*, 188: 229 (1988).
4. T. Watanabe, T. Kawada, and K. Iwai, *Proc. Soc. Exp. Biol. Med.*, 187: 370 (1988).
5. T. Watanabe, T. Kawada, and K. Iwai, *Biochem. Biophys. Res. Commun.*, 142(1): 259 (1987).
6. T. Watanabe, T. Kawada, and K. Iwai, *Agric. Biol. Chem.*, 51(1): 75 (1987).
7. K. Iwai and E. Nakatani, *Koshinryoseibun no Shokuhinkino*, Koseikan, Tokyo, 1989, p. 123.
8. J. A. LeBlanc, *Diet and Obesity* (G. A. Brag et al., eds.), Japan Sci. Soc. Press, Tokyo/ S. Karger, Basel, 1988, p. 68.
9. Y. Muto, *Syoka•Kyusyu*, Daiichi Shuppan, Tokyo, 1988, p. 120.
10. T. Kawada, K. Hagihara, and K. Iwai, *J. Nutr.*, 116: 1272 (1986).
11. C. J. K. Henry and B. Emery, *Human Nutr. Clin. Nutr.*, 40C: 165 (1986).
12. T. Yoshida, K. Yoshioka, Y. Wakabayashi, H. Nishioka, and M. Kondo, *J. Nutr. Sci. Vitaminol.*, 34: 587 (1988).
13. K. Kim, K. Ishihara, K. Inoue, T. Kawada, and T. Fushiki, *Nihon Koshinryo Kenkyukai Koen Yosisyuu*, 10: (1995).
14. T. Watanabe, *Shokuhin to Kaihatsu*, 23(6): 40 (1988).
15. T. Ariga, *Nihon Koshinryo Kenkyukai Koen Yoshisyu*, 3: (1988).
16. T. Ariga, *Shokuhin to Kaihatsu*, 28: 14 (1993).
17. K. Okano, T. Nishio, C. Fukaya, K. Yokyama, T. Nishimoto, and K. Matsmoto, *J. Med. Pharm. Soc. WAKAN-YAKU*, 9: 175 (1992).
18. K. Tazoe, *Vitamin*, 20: 97 (1960).
19. K. Yamazaki, *Aromatopia*, 13: 64 (1995).
20. *Mainichi Shinbun* September 16, 1992.
21. I. Nakajima, T. Yamanaka, S. Suzuki, and T. Maeyama, *Nihon Gan Gakkai* 47: (1988).

6

Antimicrobial and Antioxidant Properties of Spices

I. ANTIMICROBIAL PROPERTIES OF SPICES

Microorganisms play a role in many different areas of the food industry. Some are utilized effectively for producing dairy products, pickles, and other fermented products, but in most cases they cause foods to become unpalatable or to spoil.

An inhibitory action against microbial growth is generally expressed as antimicrobial action, which includes bacteriostatic or fungistatic action (preventing microbial growth and propagation), and many spices possess antimicrobial and/or antifungal properties. The antimicrobial properties of spices have been known and utilized for centuries. For example, cinnamon, cumin, and thyme were used in mummification in ancient Egypt, and spices were used in ancient India and China to preserve foods as well as for medicinal purposes. In ancient Greece and Rome, coriander was used to extend the preservation period for meat, and mint was used to prevent milk from spoiling. Infectious diseases, such as cholera and typhus, prevalent in the medieval period, were treated using spices, presumably for bactericidal as well as medical reasons. Research on the antimicrobial properties of spices began in the 1880s, and mustard, clove, and cinnamon were soon proven to have antimicrobial effects. Since the early twentieth century, research on spice extracts and the essential oils of spices has been conducted in this connection.

In this chapter, we will review various studies on the antimicrobial properties (both antibacterial and antifungal) of spices, followed by a consideration of the effectiveness of the different chemical components found in spices. Finally, we will focus on the antimicrobial properties of several pungent spices known to have strong inhibitory activity against microbes.

A. Antibacterial and Antifungal Properties of Spices

Several studies evaluated the antimicrobial activities of spice oils using a phenol coefficient, obtained by comparing the antimicrobial effect of each spice oil with that of phenol [1–3]. Martindale [1] found both clove oil and cinnamon oil to be effective in retarding microbial growth, and other researchers discovered that oregano oil and peppermint oil also had relatively strong antimicrobial effects. Maruzzella and Lichtenstein [4] tested a variety of volatile oils and terpeneless oils to observe their inhibitory activity against different bacteria using the filter paper disk method. Most volatile oils and terpeneless oils exhibited antibacterial activity for at least one of the microbes tested, and the volatile oils of cinnamon, cumin, dill weed, and thyme were found to exhibit relatively strong antibacterial activities. Of the bacteria tested, *Bacillus subtilis* was the most susceptible, and *Escherichia coli* was found to be relatively resistant to these spice essential oils. The same authors studied the antifungal activity of these volatile oils and terpeneless oils against 15 different yeasts and molds [5]. Most of the spices tested were found to possess some degree of antifungal activity. The volatile oils of thyme, cinnamon, and coriander and the terpeneless oils of anise, caraway, dill, and cinnamon exhibited greatest antifungal activity, especially against *Aspergillus* and *Streptomyces* species. Another study showed that some spice oils, including those from caraway and clove, were effective against *Staphylococcus aureus* and *E. coli* [6]. The sensitivities of various bacteria to sage, rosemary, and allspice were studied by Shelef et al. [7], who found that sage and rosemary had a wide range of antimicrobial activity compared to allspice: the minimum inhibitory concentration (MIC) of allspice was twice as high as those of sage and rosemary. A combination of rosemary and sage was found to inhibit bacterial growth more than either alone. A 0.3% concentration of either sage or rosemary was bacteriostatic, and a concentration of 0.5% had a bactericidal effect. It was also confirmed in this study that gram-positive bacteria were generally more sensitive to spices than gram-negative bacteria.

Ueda et al. evaluated the antifungal and antibacterial properties of alcohol extracts of 14 spices against several different molds and bacteria with the paper disk method [8]. As shown in Tables 6.1 and 6.2, the MIC of clove was the smallest, followed by those of cinnamon and oregano. In this study, it was also confirmed that *B. subtilis* and *S. aureus* are relatively vulnerable to spices, whereas *E. coli* is relatively resistant, and that *Aspergillus* is more susceptible than several other fungi tested. Galli et al. [9] investigated the antimicrobial spectrum of several spices against various gram-positive and gram-negative bacteria and confirmed antimicrobial activities for cinnamon, marjoram, oregano, and thyme in addition to mustard, which was found to have a wide range activity against both gram-positive and gram-negative bacteria. Miyao focused on mace, which is used often in meat products such as sausage [10]. In this study, an ethanol extract of mace mixed into sausage showed an inhibitory effect against *Bacillus megaterium*, *Acinetobacter* sp., and *Pseudomonas* sp. but exhibited only a slight effect against *Staphylococcus epidermidis*, *Proteus vulgaris*, and *Aeromonas* sp. It was also observed how long slime formation in a sausage treated with mace extract was delayed. The sausage was soaked in a solution with one of two ethanol extracts of mace (2.5 and 5.0%) for 10 seconds, then stored at both 10 and 25°C for several days. As shown in Table 6.3, slime formation was delayed by treatment with mace extract, and the preventive effect was greater at the lower temperature.

In addition to mace, Mori et al. reported that essential oils of celery, cinnamon, and cumin prevented slime formation on sausage [11]. Mace extract has been also confirmed to have antimicrobial activity against *Enterobacter aerogenes*, *Brevibacterium* and *Achromobacter* sp., *Micrococcus flavus*, *B. subtilis*, *Leuconostoc mesenteroides*, and *Lactobacillus plantarum*, but it does not retard the growth of *Serratia marcescens*.

Numerous studies have investigated the effectiveness of spices in retarding the growth of pathogenic bacteria and other toxin-producing microorganisms. In the 1940s, the effect of spices against cholera, bacillary dysentery, and tubercle bacillus became clear. Dole and Knapp, observing the antimicrobial properties of various spices against eight microorganisms including *Salmonella typhi* and *Shigella dysenteriae*, found that garlic had a strong antimicrobial effect against all eight and that both onion and clove were effective against all but *B. subtilis* [12]. *Salmonella* growth was also inhibited when this bacterium was inoculated into preenrichment media containing oregano [13]. Oregano and thyme were confirmed to be effective against *Vibrio parahemolyticus* when present in growth media in powdered form at concentrations of 0.5% [14]. The essential oils of both spices were also effective in inhibiting the growth of this bacterium at 100 μg/ml. Huhtanen tested alcohol extracts of 33 spices for inhibition of *Clostridium botu-*

TABLE 6.1 Minimum Inhibitory Concentration (%) of Spice Extracts for Bacteria

Spice	pH	B. subtilis PCI	S. aureus 209P	E. coli	S. typhimurium	S. marcescens	P. aeruginosa	P. vulgaris	P. morganii
Anise seed	7.0	4.0	2.0	4.0<	4.0<	4.0<	4.0<	4.0<	4.0<
	5.0	0.5	2.0	4.0<	4.0<	4.0<	4.0<	4.0<	4.0<
Cardamom	7.0	2.0	2.0	4.0<	4.0<	4.0<	4.0<	4.0<	4.0<
	5.0	0.1	0.5	4.0<	4.0<	4.0<	4.0<	2.0<	4.0
Caraway	7.0	4.0<	4.0	4.0<	4.0<	4.0<	4.0<	4.0<	4.0<
	5.0	4.0<	4.0<	4.0<	4.0<	4.0<	4.0<	4.0<	4.0<
Celery	7.0	4.0	1.0	4.0<	4.0<	4.0<	4.0<	4.0<	4.0<
	5.0	0.5	1.0	4.0<	4.0<	4.0<	4.0<	2.0	4.0
Cinnamon	7.0	4.0	2.0	4.0	4.0	4.0	4.0	2.0	4.0
	5.0	0.5	2.0	2.0	1.0	1.0	4.0	1.0	2.0
Cloves	7.0	1.0	1.0	1.0	1.0	1.0	2.0	1.0	1.0
	5.0	0.5	2.0	1.0	1.0	1.0	1.0	0.5	0.5
Laurel	7.0	4.0	2.0	4.0	4.0	4.0	4.0	4.0	4.0
	5.0	0.5	1.5	4.0	4.0	4.0	4.0	2.0	4.0
Mace	7.0	0.2	0.05	4.0<	4.0<	4.0<	4.0<	4.0<	4.0<
	5.0	0.1	0.5	4.0<	4.0<	4.0<	4.0<	4.0<	4.0<
Marjoram	7.0	4.0	4.0	4.0<	4.0<	4.0<	4.0<	4.0<	4.0<
	5.0	0.5	0.2	4.0<	4.0<	4.0<	4.0<	4.0<	4.0<
Oregano	7.0	1.0	1.0	2.0	2.0	4.0	4.0	2.0	4.0
	5.0	0.2	0.5	2.0	2.0	4.0	2.0	2.0	2.0
Rosemary	7.0	0.5	0.5	4.0<	4.0<	4.0<	4.0<	2.0	4.0
	5.0	0.2	0.5	4.0<	4.0<	4.0<	4.0<	0.5	4.0
Savory	7.0	2.0	2.0	4.0<	4.0<	4.0<	4.0<	4.0	4.0<
	5.0	0.2	1.0	4.0	4.0	4.0<	4.0<	4.0	4.0
Sage	7.0	1.0	0.2	4.0<	4.0<	4.0<	4.0<	4.0	4.0<
	5.0	0.2	2.0	4.0<	4.0<	4.0<	4.0<	0.5	4.0<
Thyme	7.0	2.0	1.0	4.0<	4.0<	4.0<	4.0<	4.0<	4.0<
	5.0	0.2	2.0	4.0<	4.0	4.0<	4.0<	1.0	4.0

Source: Ref. 8.

TABLE 6.2 Minimum Inhibitory Concentration (%) of Spice Extracts for Fungi

Spice	S. cerevisiae	C. parakrusei	C. krusei	P. sp.	A. oryzae
Anise seed	4.0	4.0<	4.0<	4.0	1.0
Cardamom	4.0	4.0	4.0<	4.0<	4.0<
Caraway	4.0<	4.0<	4.0<	4.0	4.0
Celery	4.0	4.0	4.0	1.0	1.0
Cinnamon	1.0	1.0	1.0	1.0	1.0
Cloves	0.5	0.5	0.5	0.5	0.2
Laurel	4.0<	4.0<	4.0<	4.0	4.0
Mace	4.0<	4.0<	4.0<	4.0<	4.0<
Marjoram	4.0<	4.0<	4.0<	4.0<	4.0<
Oregano	2.0	2.0	2.0	1.0	1.0
Rosemary	1.0	4.0<	4.0<	4.0<	4.0<
Savory	4.0	4.0	4.0	2.0	2.0
Sage	2.0	4.0<	4.0<	4.0<	4.0<
Thyme	4.0<	4.0	4.0	2.0	2.0

Source: Ref. 8.

linum, a toxic anaerobic bacterium, by determining the MIC of each spice extract [15]. Mace was the most inhibitory of the spices tested. Nutmeg, bay leaf, and black and white pepper showed strong inhibitory activity against this bacterium, while other spices including garlic, caraway, and turmeric showed little or no inhibitory activity at a concentration of 2000 ppm (Table 6.4). Some of these spices are routinely used in meat products, and it is likely that they work in these products to inhibit the growth of bacteria, depending upon their formulated quantities.

Staphylococcus aureus, which is widely found not only in the natural environment, but also on the skin and in the intestines of humans, is known to be pathogenic, causing food poisoning or suppurative conditions. Although medical treatments exist for these diseases, a new species of *S. aureus* that is tolerant to the antibiotic methicillin emerged in the 1980s (methicillin-resistant *Staphylococcus aureus*, or MRSA). Inhibitory effects of spices against MRSA have been reported [16]. Hexane extracts of 28 spices were tested for inhibition against *E. coli*, *Salmonella* sp., *S. aureus* (including MRSA), *Bacillus cereus*, and *Campylobacter*. Allspice, cinnamon, clove, marjoram, oregano, rosemary, and sage exhibited antimicrobial activities. Rosemary and sage were found to possess especially strong inhibitory activity against *S. aureus* and *B. cereus* (Table 6.5), while oregano and cinnamon showed relatively strong antimicrobial activities against *E. coli*, *Salmonella*, and *Campylobacter*.

Studies on inhibition of toxin formation and of the growth of toxin-producing fungi utilizing spices have been attracting attention recently. Farag et

TABLE 6.3 Effects of Mace on Slime Formation in Sausage

(a) Storage at 25°C

		\multicolumn{4}{c}{Days}			
		1	2	3	4
None	0%	—	++	+++	+++
Ethanol	2.5	—	+	+++	+++
Ethanol	5.0	—	+	+++	+++
Mace	2.5	—	+	+++	+++
Mace	5.0	—	±	+++	+++
Eugenol	0.1	—	+	+++	+++
Eugenol	0.2	—	+	++	+++

(b) Storage at 10°C

		Days												
		8	9	10	11	12	13	14	15	16	17	18	19	20
None	0%	—	+	+	+	++	++	+++	+++	+++	+++	+++	+++	+++
Ethanol	2.5	—	—	—	—	+	+	+	++	++	+++	+++	+++	+++
Ethanol	5.0	—	—	—	—	—	+	+	+	++	++	+++	+++	+++
Mace	2.5	—	—	—	—	+	+	+	—	++	++	+++	+++	++
Mace	5.0	—	—	—	—	—	—	—	—	+	+	++	++	++
Eugenol	0.1	—	—	—	—	—	—	—	—	—	+	+	+	+
Eugenol	0.2	—	—	—	—	—	—	—	—	—	—	—	—	+

Source: Ref. 10.

Sausages were immersed in ethanol, mace, or eugenol solution at 15°C for 10 minutes after immersion in water at 80°C for 5 seconds.

—: No slime; +: slight slime; +++: heavy slime.

TABLE 6.4 Minimum Inhibtory Concentration of Spice Extracts[a] for *Clostridium botulinum*

Common	Botanical	MIC (μg/ml)
Allspice	*Pimenta officinalis*	2000
Parsley	*Petroselinum crispum*	> 2000
Marjoram	*Marjoram hortensis*	> 500
Mustard	Blend of *Brassica hirta* and *Sinapis alba*	> 2000
Garlic	*Allium sativum*	> 2000
Celery flakes	*Apium graveolens*	> 2000
Celery seed	*Apium graveolens*	2000
Chives	*Allium schoenoprasum*	> 2000
White pepper	*Piper nigrum* (water-soaked)	125
Black pepper	*Piper nigrum* (dried fruit)	125
Sweet pepper	*Capsicum annuum*	> 2000
Paprika	*Capsicum annuum*	500
Anise	*Pimpinella anisum*	> 2000
Sage	*Salvia officinalis*	2000
Ginger	*Gingiber officinale*	2000
Caraway	*Carum carvi*	> 2000
Fennel	*Foeniculum vulgare*	> 2000
Achiote	*Bixa orellano*	31
Tarragon	*Artemesia dracunulus*	> 2000
Dill	*Anethum graveolens*	> 2000
Rosemary	*Rosmarinus officinalis*	500
Cinnamon	*Cinnamonum zeylanicum*	2000
Cloves	*Eugenia caryophyllata*	500
Red pepper	*Capsicum frutescens*	> 500
Bay leaf	*Laurus nobilis*	125
Cumin	*Cuminum cyminum*	> 2000
Oregano	*Lippia graveolens* *Oreganum vulgare*	500
Turmeric	*Circuma longa*	500
Onion	*Allium cepa*	> 2000
Thyme	*Thymus vulgaris*	500
Nutmeg	*Myristica fragrans* (seed)	125
Mace	*Myristica fragrans* (external coat)	31
Coriander seed	*Coriandrum sativum*	> 2000

[a]10% ethanol extracts. MIC values are on whole spice basis. Levels assayed were 2000, 500, 125, 31, and 8 μg/ml.
Source: Ref. 15.

TABLE 6.5 Minimum Inhibitory Concentration (%) of Hexane Extracts of Spices for Several Pathogenic Bacteria

	Escherichia coli	*Salmonella* sp.	*Staphylococcus aureus*	*Bacillus cereus*	*Campylobacter*
Allspice	10	> 10	10	10	10
Cinnamon	5	10	2.5	2.5	1.3
Clove	10	10	5	5	2.5
Marjoram	> 10	> 10	10	10	> 10
Oregano	2.5	5.	1.3	2.5	1.3
Rosemary	> 10	> 10	0.31	0.16	> 10
Sage	> 10	> 10	0.83	0.31	> 10

Source: Modified from Ref. 16.

al. [17] investigated the effectiveness of essential oil of thyme against the growth of *Aspergillus parasiticus* and its aflatoxin production in comparison with the oils of sage, clove, caraway, cumin, and rosemary. In this study, essential oil of thyme was found to inhibit mold growth and aflatoxin production at a concentration of 0.2 mg/ml, and oils of cumin, caraway, and clove were found to inhibit total flatoxin production at relatively low concentrations, although their inhibitory actions were less effective than that of thyme (Table 6.6). Bullerman [18], influenced by the fact that cinnamon bread tended not to become moldy, conducted a study to see whether cinnamon would suppress mold growth as well as whether it would prevent aflatoxin production. As shown in Table 6.7, mold growth was

TABLE 6.6 Influence of Spice Essential Oils on *Aspergillus parasiticus* Growth and Aflatoxin Production in Yeast Sucrose Medium

Oil concentration (mg/ml)	Mycelial weight (g/50 ml)	Aflatoxin concentration (μg/ml)		
		B	G	Total
Sage oil				
Control	1.61 (a)	80.4 (a)	115.1 (a)	195.5 (a)
0.2	1.41 (c)	97.9 (b)	141.6 (c)	239.5 (c)
0.4	1.25 (c)	104.3 (c)	123.0 (a)	227.3 (b)
0.6	1.15 (c)	93.5 (a)	81.7 (c)	175.2 (a)
0.8	0.80 (c)	61.5 (b)	30.7 (c)	92.2 (c)
1.0	0.71 (c)	41.6 (c)	29.7 (c)	71.3 (c)
2.0	0.20 (c)	2.2 (c)	1.3 (c)	3.5 (c)

TABLE 6.6 Continued

Oil concentration (mg/ml)	Mycelial weight (g/50 ml)	Aflatoxin concentration (μg/ml)		
		B	G	Total
Rosemary oil				
0.2	1.67 (a)	129.2 (c)	153.9 (c)	283.1 (c)
0.4	1.54 (a)	107.9 (c)	104.9 (a)	212.8 (a)
0.6	1.33 (c)	113.3 (c)	84.6 (c)	197.9 (a)
0.8	1.28 (c)	97.6 (b)	79.2 (c)	176.8 (a)
1.0	1.21 (c)	72.9 (a)	44.9 (c)	117.8 (c)
2.0	0.0	0.0	0.0	0.0
Caraway oil				
0.2	1.24 (c)	78.0 (a)	119.5 (a)	197.5 (a)
0.4	0.35 (c)	28.4 (c)	16.3 (c)	44.7 (c)
0.6	0.11 (c)	1.1 (c)	0.5 (c)	1.6 (c)
0.8	0.0	0	0	0
1.0	0.0	0.0	0.0	0.0
2.0	0.0	0.0	0.0	0.0
Clove oil				
0.2	1.20 (c)	87.6 (a)	110.0 (a)	197.6 (a)
0.4	0.12 (c)	0.4 (c)	0.3 (c)	0.7 (c)
0.6	0.0	0.0	0.0	0.0
0.8	0.0	0.0	0.0	0.0
1.0	0.0	0.0	0.0	0.0
2.0	0.0	0.0	0.0	0.0
Cumin oil				
0.2	0.91 (c)	50.0 (c)	66.9 (c)	116.9 (c)
0.4	0.09 (c)	1.2 (c)	0.9 (c)	2.1 (c)
0.6	0.0	0.0	0.0	0.0
0.8	0.0	0.0	0.0	0.0
1.0	0.0	0.0	0.0	0.0
2.0	0.0	0.0	0.0	0.0
Thyme oil				
0.2	0.25 (c)	2.4 (c)	3.6 (c)	6.0 (c)
0.4	0.0	0.0	0.0	0.0
0.6	0.0	0.0	0.0	0.0
0.8	0.0	0.0	0.0	0.0
1.0	0.0	0.0	0.0	0.0
2.0	0.0	0.0	0.0	0.0

Numbers in a column followed by the same letter are not significantly different at $P = 0.01$.
Source: Ref. 17.

TABLE 6.7 Effect of Ground Cinnamon on Growth and Total Aflatoxin Production by *Aspergillus parasiticus* NRRL 2999 and NRRL 3000 in Yeast-Extract Broth after 10 Days at 25°C

Level of cinnamon (%)	Strain NRRL 2999				Stain NRRL 3000			
	Mycelia		Aflatoxins		Mycelia		Aflatoxins	
	mg	Inhibition (%)	µg/ml	Inhibition (%)	mg	Inhibition (%)	µg/ml	Inhibition (%)
Control	2301	—	356	—	1896	—	292	—
0.02	1943	16	267	25	1959	(+3)	232	21
0.2	1768	23	148	58	1557	18	49	83
2.0	1589	31	11	97	1658	13	2	99
20.0	ND	100	ND	100	100	95	0.3	99.9

ND = None detected.
Source: Ref. 18.

suppressed concomitantly with the addition of cinnamon. The addition of 0.02% cinnamon powder suppressed aflatoxin production 21–25%, and aflatoxin production was almost completely inhibited at cinnamon concentration of 2.0%. Cinnamon bread usually contains 0.5–1.0% cinnamon, indicating that mold growth and aflatoxin production should be almost completely inhibited. Ethanol extract of cinnamon was confirmed to possess strong inhibitory activity for both mold and aflatoxin production (Table 6.8). It is interesting to note that the addition of 0.02% ethanol extract of cinnamon inhibited production of aflatoxin B_1, B_2, G_1, and G_2 74–89%, although mold growth was not inhibited at this concentration. This would indicate that higher spice concentrations may be necessary to inhibit mold growth than to suppress aflatoxin production. The effect of 29 spices (not including cinnamon) on three toxigenic *Aspergillus* sp. (*A. flavus*, *A. ochraceus*, and *A. versicolor*) was observed by Hitokoto et al. [19]. Their study showed that clove, star anise, and allspice caused complete inhibition of all three *Aspergillus* strains.

B. Antibacterial and Antifungal Properties of Chemical Components of Spices

In addition to the above research on spices and their essential oils, the antimicrobial activity of various constituents found in these spices has also been studied.

Eugenol, a major component of clove and allspice (also contained in cinnamon as a secondary component), was studied by Miyao [10]. This compound exerted complete inhibition against *Acinetobacter* sp. and yeast at a concentration

TABLE 6.8 Effect of Alcohol Extract of Cinnamon on Growth and Total Aflatoxin Production by *Aspergillus parasiticus* NRRL 2999 in Glucose Ammonium Nitrate Broth after 10 Days at 25°C

Treatment (%)	Mycelial weight		Aflatoxins/ml						
			B_1		B_2		G_1		G_2
	mg	Inhibition (%)	µg/ml	Inhibition (%)	µg/ml	Inhibition (%)	µg/ml	Inhibition (%)	µg/ml	Inhibition (%)
Control	534	—	5.52	—	1.01	—	4.11	—	1.01	—
0.02	564	—	1.44	74	0.16	84	0.90	78	0.12	89
0.20	466	13	0.83	98	0.007	99+	0.03	99	0.006	99
1.00	126	76	0.008	99+	0.001	99+	0.01	99+	0.001	99+
2.00	108	80	0.007	99+	0.001	99+	0.006	99+	0.001	99+

Source: Ref. 18.

TABLE 6.9 Inhibitory Effects of Eugenol on Growth and Toxin Production of Toxigenic Fungi in Appropriate Broth for Toxin Production

Eugenol (μg/ml)	A. flavus		A. ochraceus		A. versicolor	
	Mycelia (mg)	Aflatoxin B_1 (μg/ml)	Mycelia (mg)	Ochratoxin A (μg/ml)	Mycelia (mg)	Sterigmatocystin (μg/ml)
500	0 (100)[a]	0 (100)	0 (100)	0 (100)	0 (100)	0 (100)
250	0 (100)	0 (100)	6 (98)	0.5 (83)	0 (100)	0 (100)
125	110 (40)	0 (100)	248 (32)	0.7 (76)	427 (27)	0.8 (95)
62.5	148 (19)	2.5 (83)	335 (8)	3.0 (0)	555 (5)	4.5 (70)
31.2	182 (14)	3.5 (76)	399 (0)	2.5 (17)	563 (4)	6.0 (60)
Control	183	15.0	365	3.0	585	15.0

[a]Numbers in parentheses indicate percent inhibition.
Source: Ref. 19.

of 200 ppm and against *Bacillus megaterium* and *Pseudomonas* sp. at 800 ppm. It has also been reported that both *A. flavus* and *A. versicolor* were completely inhibited by eugenol at a concentration of 250 ppm [19] (Table 6.9). Anetol, the major volatile compound of anise seed, and thymol, which is contained in thyme, also showed inhibitory activities against these *Aspergillus* species and against aflatoxin production, although their activities were lower than eugenol. As shown in Tables 6.10 and 6.11, anetol and thymol inhibited the growth of molds completely at 0.2% and 200 ppm, respectively.

Several studies reported on the antimicrobial effects of highly volatile compounds found in spice essential oils. Katayama et al. studied the antimicrobial properties of 43 kinds of phenol and terpene compounds with low molecular weights against six kinds of bacteria, including *E. coli, Salmonella enteritidis*, S.

TABLE 6.10 Inhibitory Effects of Anethol on Growth and Toxin Production of Toxigenic Fungi in Appropriate Broth for Toxin Production

Anethol (mg/ml)	A. flavus		A. ochraceus		A. versicolor	
	Mycelia (mg)	Aflatoxin B_1 (μg/ml)	Mycelia (mg)	Ochratoxin A (μg/ml)	Mycelia (mg)	Sterigmatocystin (μg/ml)
2	0 (100)[a]	0 (100)	0 (100)	0 (100)	0 (100)	0 (100)
1	270 (0)	15.0 (0)	439 (0)	0.8 (73)	253 (57)	1.0 (93)
0.5	246 (0)	15.0 (0)	614 (0)	1.0 (67)	426 (28)	6.0 (60)
0.25	258 (0)	15.0 (0)	607 (0)	1.5 (50)	511 (13)	15.0 (0)
Control	183	15.0	365	3.0	585	15.0

[a]Numbers in parentheses indicate percent inhibition.
Source: Ref. 19.

TABLE 6.11 Inhibitory Effects of Thymol on Growth and Toxin Production of Toxigenic Fungi in Appropriate Broth for Toxin Production

Thymol (μg/ml)	A. flavus		A. ochraceus		A. versicolor	
	Mycelia (mg)	Aflatoxin B_1 (μg/ml)	Mycelia (mg)	Ochratoxin A (μg/ml)	Mycelia (mg)	Sterigmatocystin (μg/ml)
400	0 (100)[a]	0 (100)	0 (100)	0 (100)	0 (100)	0 (100)
200	51 (72)	0.3 (98)	12 (96)	0 (100)	0 (100)	0 (100)
100	138 (25)	15.0 (0)	155 (58)	5.0 (0)	230 (61)	0.5 (97)
50	237 (0)	15.0 (0)	414 (0)	5.0 (0)	385 (34)	2.5 (83)
25	NT[b]	NT	NT	NT	597 (0)	15.0 (0)
Control	183	15.0	365	3.0	585	15.0

[a] Numbers in parentheses indicate percent inhibition.
[b] NT = not tested.
Source: Ref. 19.

aureus, and others [20]. As shown in Table 6.12, of these 43 compounds, thymol (a major compound of thyme and rosemary), borneol (contained in sage and rosemary), isoborneol (contained in thyme), vanillin, salicylic aldehyde, carvacrol, and eugenol all showed antibacterial activity against most of the microbes tested even at 1000-fold dilution.

Kurita et al. [21] examined the antifungal activities of perillaldehyde, cinnamaldehyde, and other chemical compounds. The antifungal activities of each compound were evaluated based on duration of growth inhibition for 18 kinds of fungi. As shown in Table 6.13, cinnamaldehyde exhibited inhibitory effects on most of fungi tested at a concentration of 0.66 mM. Perillaldehyde also showed antifungal activity against some of these fungi, although its anti-fungal activity was lower than that of cinnamaldehyde. Cinnamaldehyde also showed strong inhibitory effects on *Saccharomyces cerevisiae* as well as on *Penicillium* and *Aspergillus* sp. [8].

As explained so far, chemical compounds with a hydroxyl group (−OH) or an aldehyde group (−CHO) tend to exhibit strong antimicrobial activity. It is well known that a hydroxyl group can form hydrogen bonds with the active site of an enzyme, resulting in its deactivation. The growth inhibition by the aldehyde group is considered to be due partially to their reactions with sulfhydryl groups involved in microbial growth [17].

C. Antibacterial and Antifungal Properties of Pungent Spices

Most pungent spices are said to have relatively strong antimicrobial properties, and this belief has been supported by numerous studies. Most antimicrobial

TABLE 6.12 Antibacterial Activity of Terpene Volatile Alcohols and Aldehydes

Compounds	B. subtilis	E. coli	S. enteritidis	S. aureus	P. morganii	P. aeruginosa
Linalylacetade	20	10	20	20	20	10
Terpenylacelate	100	—	—	100	10	—
n-Hexylalcohol	100	100	100	20	100	100
n-Octylalcohol	200	200	200	200	200	200
Furfurylalcohol	10	—	10	20	20	100
Benzylalcohol	20	20	20	20	20	200
σ-Terpineol	20	20	20	20	20	20
Citronellol	200	20	100	100	100	20
Geraniol	1000	200	200	200	200	200
Linalool	10	20	20	10	10	20
Eugenol	<2000	<2000	<2000	<2000	<2000	<2000
Thymol	1000	<2000	2000	1000	1000	<2000
o-Cresol	100	100	100	100	1000	1000
m-Cresol	100	100	100	100	100	100
p-Cresol	100	100	100	20	100	100
Carvacrol	2000	200	1000	2000	<2000	<2000
Isoborneol	2000	—	1000	1000	1000	1000
Vanilline	2000	2000	2000	2000	<2000	200
Isovanilline	100	100	100	100	100	200
Salicylic aldehyde	2000	<2000	<2000	<2000	2000	1000
Furfural	20	20	20	20	20	100
Cinnamic aldehyde	200	20	100	200	100	20
Anisaldehyde	20	100	200	20	1000	1000
Citronellal	100	10	100	100	20	20
Citral	200	100	100	100	100	20
Perillaldehyde	100	20	20	20	20	10
Carvone	100	20	20	20	20	20
Pseudoionone	200	20	20	200	200	1000
Comphene	10	10	10	—	—	20
Fenchon	20	10	20	20	20	20
Benzaldehyde	20	100	100	20	100	100
Acetaldehyde	20	20	20	20	20	20
Safrol	100	—	100	100	100	100
Isosafrol	200	10	1000	200	200	100
1:8-Cineol	—	20	10	10	10	10
Ascaridol	10	10	10	10	10	20
α-Pinene	10	10	10	10	10	10
β-Pinene	100	20	20	20	100	100
Terpinolene	20	20	100	20	20	200
d-Limonene	200	10	10	10	10	100
σ-Phellandrene	10	20	20	10	10	10
p-Cymene	—	—	—	—	—	10
Lemon oil	100	—	10	10	10	—

Numerical value indicates dilute magnification that showed inhibitory activity against each microbe.
Source: Ref. 20.

TABLE 6.13 Antifungal Activity of Aldehyde Compounds

	Duration of growth inhibition (day)							
	Cinnamaldehyde (mM)		Perillaldehyde (mM)		Citral (mM)		Citronellal (mM)	
Fungus	0.33	0.66	0.33	0.66	0.33	0.66	0.33	0.66
Fonsecaea pedrosoi (Tsuchiya)	0	2	0	1	0	0	0	0
Cladosporium bantianum (1169)	0	>20	0	3	0	1	0	0
Trichophyton rubrum (Hagiwara)	3	>20	0	7	1	>20	0	0
Trichophyton mentagrophytes	>20	>20	2	4	1	>20	0	0
Trichophyton violaceum	>20	>20	10	>20	7	>20	0	0
Microsporium gypseum (N-198)	>20	>20	2	12	3	>20	0	0
Histoplasma capsulatum	>20	>20	6	13	2	>20	0	0
Blastomyces dermatitidis	>20	>20	4	10	4	>20	0	0
Sporothrix schenckii (sp-56)	1	>20	0	1	0	2	0	0
Ceratocystic ips (1687)	4	>20	0	2	0	0	0	0
Penicillium decumbens	0	>20	0	0	0	1	0	0
Penicillium rugulosum	1	>20	0	0	1	3	0	0
Penicillium frequentans	0	1	0	1	0	1	0	0
Aspergillus niger	0	1	0	0	0	1	0	0
Aspergillus fumigatus	0	>20	0	0	1	2	0	0
Aspergillus nidurans	1	>20	0	0	0	0	0	0
Candida albicans (N 500)	0	>20	0	0	0	0	0	0
Cryptococcus neoformans	6	>20	0	1	0	2	0	0

Source: Ref. 21.

components of pungent spice are found in volatile components of their essential oils, except in the case of red and black/white pepper. The inhibitory activities of red and black/white pepper, which are not contained in their essential oils, are also thought to be due to their pungent principles, but their activities are weaker than those in volatile pungent compounds. In this section we consider the antimicrobial effects of individual pungent spices.

1. Mustard and Wasabi

Mustard belongs to the Cruciferae family, and its seeds are utilized as a spice. Pungent mustard oil is obtained by mixing milled seeds with water. Mustard oil is known to be effective in preventing the growth of a wide variety of microorganisms. The major antimicrobial principle of black mustard (*Brassica nigra*) is allylisothiocyanate, which is produced by the action of the glucosinolase myrosinase on the glycoside sinigrine, and that of white mustard (*Sinapsis alba*) is *p*-hydroxybenzyl-isothiocyanate, produced by the action of the same glucosinolase on the glycoside sinalpine. Wasabi (*Wasabia japonica*) and horseradish, which also belong to the Cruciferae family, produce allyl-isothiocyanate by the action of myrosinase when their plant tissues are disrupted, and they also show strong antimicrobial activities. Many different kinds of isothiocyanate compounds exist in both types of mustard, wasabi, and horseradish as forms of glycosides, and these compounds are also said to have antimicrobial activities, although their effects are relatively small compared to the major pungent compounds listed in the above. Most of these compounds are volatile and lose their effectiveness quickly, although their antimicrobial activities are in general, stronger than those of ginger and red pepper.

Many studies have reported on how spice oils or pungent compounds inhibit the growth of many types of microorganisms. White mustard soaked in water was confirmed to be effective against various fungi existing on human skin. Miyamoto studied the antimicrobial properties of an ethanol extract of wasabi using *E. coli* [22]. Both absorbency and sugar consumption were used as indicators of antimicrobial activities of this extract. As shown in Table 6.14, as microbes multiply, the culture solution becomes turbid, has higher absorbency, and the sugar value increases as the energy of microbes is consumed. Miyamoto found that the antimicrobial property of this extract against *E. coli* was enhanced when added in increased amounts.

The antimicrobial activities of mustard and its major pungent compound,

TABLE 6.14 Effect of Wasabi Powder on Growth of *Escherichia coli*

Cultivating time (hr)	0% wasabi		0.2% wasabi		0.4% wasabi	
	OD_{600}	Sugar consumption (mg/ml)	OD_{600}	Sugar consumption (mg/ml)	OD_{600}	Sugar consumption (mg/ml)
0	0.00	0.0	0.00	0.0	0.00	0.0
6	0.46	0.5	0.00	0.5	0.00	0.0
12	0.93	2.2	0.00	0.1	0.00	0.0
18	0.97	2.6	0.06	0.0	0.00	0.0
24	1.03	3.2	0.64	1.6	0.00	0.5
30	—	—	0.85	2.2	0.13	0.8
48	1.41	4.6	1.06	2.9	0.89	2.3
72	1.54	6.1	1.38	5.2	1.15	4.6
96	1.70	8.2	1.49	5.8	1.30	5.5

Source: Ref. 22.

allyl-isothiocyanate (also the major pungent principle of wasabi), against the growth of several kinds of bacteria were studied [23]. Both allyl-isothiocyanate and mustard were shown to inhibit the growth of *E. coli* and *S. aureus*, and both showed an even stronger inhibitory effect against *Pseudomonas* sp. Isshiki et al. [24] evaluated the effectiveness of allyl-isothiocyanate and its vapor on various microbes. In this experiment, a filter paper to which 100 μl of allyl-isothiocyanate–corn oil mixture was added was placed inside of top cover of the petri dish, then the plate containing each microbe was incubated. The effect of allyl-isothiocyanate and its vapor was examined by measuring minimum inhibitory dose (MID) and MIC in headspace gas. As shown in Table 6.15, 31–470 μg/dish (equivalent of 16–110 ng/ml of allyl-isothiocyanate vapor) suppressed the growth of bacteria and yeasts. Allyl-isothiocyanate and its vapor inhibited molds and yeast more effectively than it did bacteria. The antimicrobial effect is thought to be due to the reaction of thiocyanate ($-N=C=S$) with $-SH$ in the cytoplasm or cytoplasmic membrane.

Recently, some products making use of the antimicrobial properties of allyl-isothiocyanate have been commercialized. These are now being used to preserve, prevent mold, decrease the number of microbes, and prevent insect infestation in a number of food products, including breads, pasta, lettuce, grains, dried mushrooms, and fruits.

The effect of allyl-isothiocyanate was compared with those of other spice components including carvacrol (thyme) and salicylaldehyde (cinnamon). Allyl-isothiocyanate was found to retard the growth of most kinds of microorganisms tested, but its effect on gram-positive bacteria was relatively low, while salicylaldehyde and carvacrol were confirmed to be effective for both gram-negative and gram-positive bacteria. The effectiveness of allyl-isothiocyanate against some kinds of molds is enhanced synergistically when it is combined with other compounds.

2. Garlic

Garlic has long been used in foods as a preservative as well as to give its typical flavor. Cavallito and Bailey [25] and Chain [26] tested the antimicrobial activities of garlic and its constituents using cylinder plate methods. Natural diallyl sulfides and diallyl polysulfides, which are major flavor compounds of garlic, did not show antimicrobial activity, but allicin demonstrated inhibitory activity for the growth of both gram-positive and gram-negative bacteria.

Noda et al. [27] investigated the antifungal activity of garlic against typical fungi (*Saccharomyces cerevisiae* and *Aspergillus oryzae*) and found that it was drastically reduced when it was heated (which deactivates the enzyme leading to the production of allicin) before it was used. Garlic juice at a concentration of 0.5% was found to inactivate typhoid bacillus completely in 5 minutes, and it inhibited the growth of almost all kinds of microorganisms at a concentration of

TABLE 6.15 Antimicrobial Activity of Allyl Isothiocyanate Vapor

Organism	MID (μg/dish)	MIC (ng/ml)
Bacteria		
Bacillus subtilis IF-13722	420	110
Bacillus cereus IFO-13494	360	90
Staphylococcus aureus IFO-12732	420	110
Staphylococcus epidermidis IFO-12993	420	110
Escherichia coli JCM-1649	110	34
Salmonella typhimurium ATCC-14028	210	54
Salmonella enteritidis JCM-1891	420	110
Vibrio parahaemolyticus IFO-12711	210	54
Pseudomonas aeruginosa IFO-13275	210	54
Yeasts		
Saccharomyces cerevisiae NFRI-3066	62	22
Hansenula anomala NFRI-3717	124	37
Torulaspora delbreuckii NFRI-3811	50	18
Zygosaccharomyces rouxii NFRI-3447	31	16
Candida tropicalis NFRI-4040	62	22
Candida albicans IFO-1061	62	22
Molds		
Aspergillus niger ATCC-6275	124	37
Aspergillus flavus NFRI-1157	124	37
Penicillium islandicum NFRI-1156	62	22
Penicillium citrinum NFRI-1019	62	22
Penicillium chrysogenum IFO-6223	250	62
Fusarium oxysporum NFRI-1011	62	22
Fusarium graminearum NFRI-1233	31	16
Fusarium solani IFO-9425	110	34
Alternaria alternata IFO-4026	62	22
Mucor racemosus IFO-6745	250	62

Growth of bacteria and molds was observed for 7 days.
MID = minimum inhibitory dose/petri dish (9 cm dia.); MIC = minimum inhibitory concentration in headspace gas.
Source: Ref. 24.

3.0%. However, garlic juice was also reported to sometimes promote the growth of E. coli [27]. In this experiment, several concentrations of garlic juice were used (Fig. 6.1). No growth of E. coli was observed at 750-fold dilution, but E. coli was observed in the range of 15,000- to 150,000-fold dilution. Further study disclosed that scordinine, a constituent of garlic, has the ability to promote propagation of this E. coli.

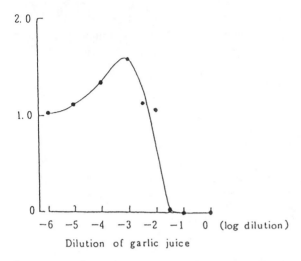

FIGURE 6.1 Growth response of *Escherichia coli* cultured in a media containing different concentrations of garlic juice. The average numbers of colony formed without garlic juice are regarded as a standard (1.0). (From Ref. 27.)

3. Red pepper

Red pepper is known to prevent the growth of molds and bacteria to some extent. Capsaicin, the major pungent compound in red pepper, is responsible for its antimicrobial property. Red pepper is also said to be effective as an anticholera agent.

4. Ginger

The essential oil of ginger has been shown to inhibit both cholera and typhoid. Gingerone and gingerol in particular, the major pungent components of this spice, show strong inhibitory effects against these pathogenic bacteria.

II. ANTIOXIDANT PROPERTIES OF SPICES

A. History

Foods deteriorate gradually during storage for various reasons, one of which is the oxidation of oil or fat contained in foods. The use of antioxidants could prolong the shelf life of many kinds of foods and make it possible to market many new food products.

Most foods contain some fat, which is an important constituent for many reasons. Fats, however, are known to react with oxygen in the air to generate peroxides, which are further oxidized and decompose into low molecular alcohol and aldehyde compounds, resulting in rancidity. These "free radicals" are also thought to work in the human body to damage DNA and promote cancer and aging. β-Carotene, found in red and green vegetables, is one of the antioxidants studied in this connection.

The problem of oxidation in packaged foods is sometimes addressed by replacing the oxygen in the container with an inert gas such as nitrogen, or sometimes a deoxidizer is put in the container, but the most common method is to employ antioxidants in the food. Two well-known synthetic antioxidants, butylated hydroxyanisol (BHA) and butylated hydroxytoluene (BHT), have been used in a wide variety of foods. However, both are very volatile and easily decomposed at high temperature (i.e., not appropriate for fried foods). There is also concern about possible toxicity to both the liver and the lungs. Tocopherol, a natural antioxidant, has been widely used in the food industry, but its antioxidant effect is inferior to the synthetic antioxidants.

B. Antioxidant Properties of Spice

Studies on the antioxidant properties of several kinds of spice in the 1930s found that some spices retarded the generation of peroxide in peanuts oil and inhibited rancidity in meats. The first patent for spice antioxidants, approved in 1938, specified that spice oils such as clove oil could prevent the oxidation of cooking oils [28]. Dubois and Tressler, in a study on the antioxidant effects of various spices for the purpose of preventing oxidation of minced pork during frozen storage, found that nutmeg, mace, black pepper, sage, and ginger were effective [29]. Chipault et al. evaluated the antioxidant properties of 72 kinds of spices, their petroleum ethers, and their alcohol-soluble fractions, and found 32 spices to retard the oxidation of lard [30]. In their study, the antioxidant activities of ground sage and rosemary were particularly strong, and oregano, thyme, turmeric, and nutmeg possessed relatively strong antioxidant activities. The antioxidant effects of several spices in foods containing fats or oils are summarized in Table 6.16. In the table, antioxidant activity is expressed as an antioxidant index, which is the ratio of the stability of each food containing spice to that containing no spice.

A number of studies on the antioxidant properties of spices were undertaken in the 1970s. Watanabe and Ayano examined the antioxidative activities of 11 different ground spices and their water-soluble and ethanol-soluble fractions by the active oxygen method (AOM) [31]. As shown in Table 6.17, sage and rosemary extended the AOM hour to reach a peroxide value of 100 meq/kg more than 10 times at concentrations of 0.25%, but almost no effects were found for black pepper, red pepper, and allspice. As for solvent extracts, almost all antioxidant activities appeared to occur in ethanol-soluble fractions. The only exception was clove, which seemed to contain effective compounds in both the water-soluble and ethanol-soluble fractions.

TABLE 6.16 Antioxidant Index[a] of Ground Spices in Various Foods

	Lard 0.2%	Pie crust 0.2%	O/W emulsion 0.1%	Minced pork −5°C 0.25%	Minced pork −15°C 0.25%	Mayonnaise 0.2%	Salad dressing 37°C 1.0%	Salad dressing 63°C 1.0%
Allspice	1.8	1.1	16.7	5.3	10.0	1.4	1.1	1.2
Clove	1.8	1.3	85.8	5.3	10.0	2.0	2.0	1.2
Oregano	3.8	2.7	7.9	7.2	3.7	8.5	2.6	2.4
Rosemary	17.6	4.1	10.2	5.3	10.0	2.2	—	—
Sage	14.2	2.7	7.8	5.3	10.0	2.4	2.2	2.2
Thyme	3.0	1.9	6.8	6.0	3.2	1.8	—	—

[a]Antioxidant index is the ratio of the stability of the sample containing spice to that of the sample to which spice is not added.

Saito et al. studied the antioxidant properties of a wide variety of ground spices and their petroleum ether–soluble and –insoluble fractions compared with those of tocopherol and BHA [32]. As shown in Table 6.18, many leaves (edible herbs) were found to strongly inhibit the oxidation of lard. Rosemary and sage exhibited stronger activities than BHA at the same concentration. Many other spices, including mace, thyme, marjoram, oregano, clove, and ginger, were more effective than tocopherol. Nitta reported that rosemary retarded rancidity the most, followed by sage, ginger, nutmeg, thyme, clove, mace, and oregano [33]. Interestingly, black pepper exhibited more remarkable antioxidative activity when used in mayonnaise or salad dressing than did rosemary.

In addition to the antioxidant properties of each individual spice, it has been confirmed that there are synergistic effects for spice-spice and spice-antioxidant combinations. The antioxidant activities of spices are known to be enhanced synergistically when tocopherol or sodium ascorbate are used. Table 6.17 shows that water extracts of many spices exhibited synergistic effects with α-tocopherol.

From the numerous studies on the antioxidative properties of spices, it can be concluded that sage and rosemary are by far the most effective in retarding the oxidation of fats or oils and that other leafy spices such as oregano and thyme have stronger antioxidant activities than most other spices.

C. Antioxidant Components of Spices

As indicated above, studies conducted over many years have confirmed the antioxidant properties of certain spices. Recent studies have focused on the isolation and identification of the chemical substances responsible. Many spices belonging to the Labiatae family show particularly strong antioxidant activity. A relatively

TABLE 6.17 Antioxidant Activities of Distilled Water–Soluble and Ethanol-Soluble Fractions of Ground Spices in Lard

Spice	Ground spice (AOM hr)	Distilled water–soluble fraction			Ethanol-solution fraction		
		α-Tocopherol (AOM hr)		Actual concentration (mg%)	α-Tocopherol (AOM hr)		Actual concentration (mg%)
		0 mg%	10 mg%		0 mg%	10 mg%	
Control	5.7	5.7	21.9	—	5.7	21.9	—
Allspice	11.8	9.7	32.8	27.5	8.7	15.5	26.7
Black pepper	7.7	5.8	28.9	16.8	6.4	24.7	18.6
Capsicum	8.0	6.8	31.7	56.5	7.0	21.8	50.1
Clove	19.6	13.9	33.4	53.6	18.9	27.5	52.8
Ginger	14.7	6.7	30.8	37.9	11.7	24.7	14.6
Mace	24.0	6.3	27.1	34.9	21.7	34.6	96.9
Nutmeg	21.7	6.1	27.3	15.7	18.5	21.7	35.2
Rosemary	67.0	8.2	29.7	46.7	58.5	58.1	41.9
Sage	54.8	6.9	26.6	47.1	42.1	42.7	33.6
Turmeric	16.8	7.0	27.7	17.1	12.4	24.5	19.2

Ground spices were incorporated at concentration of 0.25%. The distilled water–soluble and ethanol-soluble fractions were incorporated at concentration equivalent to 0.25% of the ground spice.
AOM hours: hours to reach peroxide value of 100 meq/kg.
Source: 31.

TABLE 6.18 Antioxidative Activities of Various Spices Against Lard

Spice	Ground spice, POV (mEq/kg)	Petroleum ether–soluble fraction, POV (mEq/kg)	Petroleum ether–insoluble fraction, POV (mEq/kg)
Leaf-type spices			
Basil	254.8	453.1	55.6
Bay leaf	345.8	366.9	51.4
Marjoram	23.9	5.1	28.7
Oregano	38.1	21.9	316.0
Rosemary	3.4	6.2	6.2
Sage	2.9	5.0	5.0
Tarragon	202.0	503.0	46.2
Thyme	18.3	7.3	22.0
Other spices			
Allspice	298.0	37.4	494.9
Cardamom	423.8	711.8	458.6
Black pepper	364.5	31.3	486.5
Red pepper	108.3	369.1	46.2
Japanese pepper	430.2	485.1	340.7
Cinnamon	324.0	36.4	448.9
Clove	22.6	33.8	12.8
Ginger	40.9	24.5	35.5
Turmeric	399.3	430.6	293.7
Anise seed	341.0	53.9	462.3
Caraway	396.3	589.1	293.7
Celery seed	347.2	54.5	430.0
Coriander	364.8	64.8	528.6
Cumin	600.0	59.8	479.4
Dill seed	355.2	364.0	429.7
Fennel seed	331.9	104.9	529.0
Mace	13.7	29.0	11.3
Nutmeg	205.6	31.1	66.7
Control (no additive)	356.5		
BHA	12.2		
Tocopherol	58.4		

Concentration added was 0.02%.
Source: Ref. 32.

large number of studies have been conducted on the antioxidant components of rosemary. After the active antioxidative compound carnosol was isolated from this spice by Brieskorn et al. [34], Nakatani et al. [35–37] isolated three more antioxidant substances from rosemary—all phenol diterpene compounds—and determined the structural formula of each. Rosmanol, and isomers, epirosmanol and isorosmanol are shown in Figure 6.2. Rosmanol, an odorless and colorless compound, demonstrated antioxidant activity more than four times greater than the synthetic antioxidants BHA and BHT (Fig. 6.3). Epirosmanol and isorosmanol possess antioxidative activities to almost the same degree. Both rosmanol and epirosmanol are also found in sage. More recent studies [38,39] have identified two more chemical constituents of rosemary: rosmaridiphenol, a diphenolic diterpene, and rosmariquinone. Both of these exhibited antioxidant activities superior to BHA but slightly less effective than BHT (Fig. 6.4).

Several active compounds have been isolated from oregano, another strong antioxidant spice [40,41]. Figure 6.5 shows the chemical structures for phenolic glucoside (4-phenyl-β-D-glucopyranoside), caffeic acid, protocatechuic acid, rosmarinic acid, and 2-caffeoyloxy-3-phenylpropionic acid. These compounds were more effective antioxidants than α-tocopherol, and 4-phenyl-β-D-glucopyranoside and 2-caffeoyloxy-3-phenylpropionic acid possessed antioxidative activities comparable to BHA. Phenolic glucoside is characterized by its solubility in water, and its use would be expected to be beneficial for some cases. There are

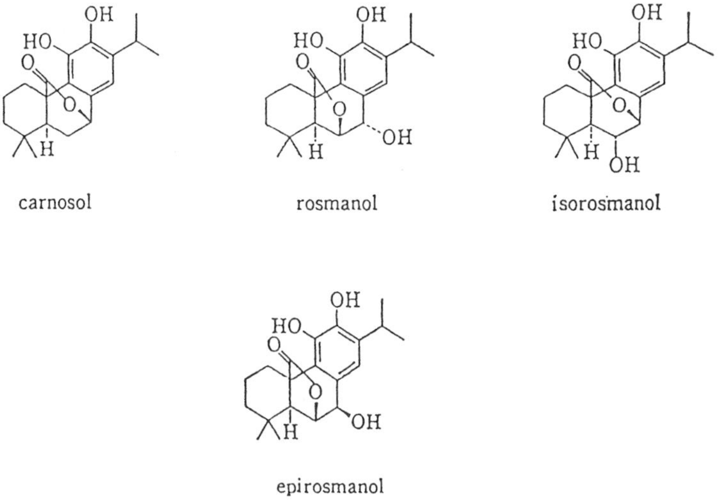

FIGURE 6.2 Structures of antioxidative compounds isolated from rosemary. (From Refs. 35–37.)

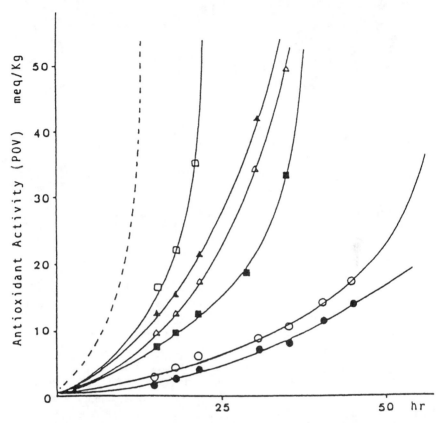

FIGURE 6.3 Antioxidative activities of rosmanol and carnosol evaluated with AOM. ●–●, Rosmanol 0.02%; ○–○, rosmanol 0.01%; ■–■, carnosol 0.01%; □–□, α-tocopherol 0.02%; ▲–▲, BHT 0.02%; △–△, BHA 0.02%; – – –, control. (From Ref. 35.)

also some studies on the antioxidant activities of nutmeg and mace [32,33,42]. Mace is said to be more effective than nutmeg in retarding the autoxidation of lard, although the chemical components contained in each are almost same. As shown in Table 6.19, strong antioxidant properties were found for both petroleum ether–in soluble and –soluble fractions of mace. The ether-soluble fraction of this spice was refined to an odorless and colorless crystal, myristphenone, and the structural formula of this crystal was determined. This chemical compound showed antioxidant activity two to four times as strong as BHA in lard and four times as strong as BHA in soybean oil. It was also proved that myristphenone retarded autoxidation and off-flavor development in food products. Sesame oil has been known for a

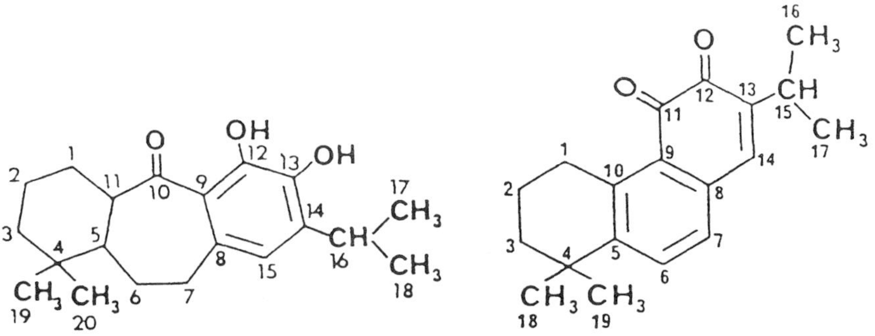

FIGURE 6.4 Structures of rosmaridiphenol and rosmariquinone. (From Refs. 38, 39.)

long time to have strong antioxidant properties and is known to be the most stable of a variety of vegetable oils in terms of autoxidation. The natural antioxidant γ-tocopherol occurs in sesame seed (*Sesamum indicum*) in relatively large quantities (approximately 0.02%). Several other kinds of spices contain tocopherol, and it was thought at first that tocopherol accounted for all of the antioxidant properties of such spices including sesame. Subsequent studies have revealed several other compounds contained in sesame oil that have antioxidant properties [43–45]: sesamol, sesamol dimer, syringic acid, ferulic acid, and lignin compounds, including sesaminol and sesamolinol (Fig. 6.6). Sesaminol, with much higher antioxidant activity than γ-tocopherol, occurs in sesame seed in amounts four times that of γ-tocopherol. It can, therefore, be assumed that the synergistic effects of these lignin compounds with γ-tocopherol account for the strong antioxidant property of sesame oil. Fukuda et al. explained that the active compound sesaminol was generated from sesamolin, which has no antioxidant activity, during the refining of this oil [45].

Antioxidant components contained in thyme, ginger, red pepper, turmeric, and other spices have been identified to some extent. Fujio et al. tested the antioxidative effects of several essential oils and their compounds, as well as Welsh onion, garlic, and ginger on dehydrated pork [46]. Eugenol and thymol, which are the major compounds found in the essential oils of clove and thyme, respectively, retarded the increase in peroxide value (POV) of dehydrated pork the most, while linalool and cineol, which are contained in many spices such as ginger, nutmeg, cinnamon, and sage, did not show marked antioxidant properties (Fig. 6.7). The addition of allylsulfide from Welsh onion or allyldisulfide from garlic prior to the dehydration of pork apparently retarded oxidation. Shogaol and zingerone, both pungent compounds found in ginger, exhibited strong antioxidant

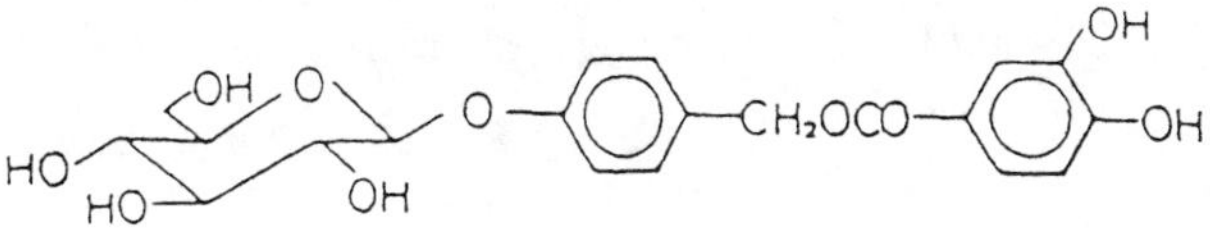

1

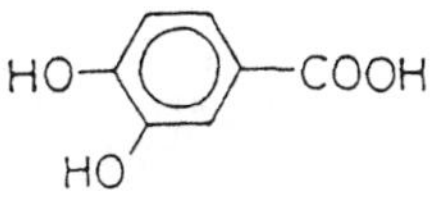

2

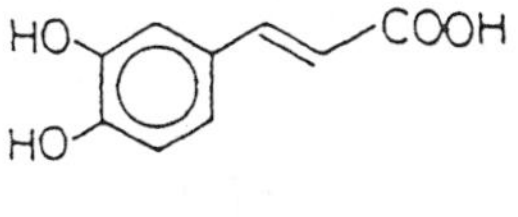

3

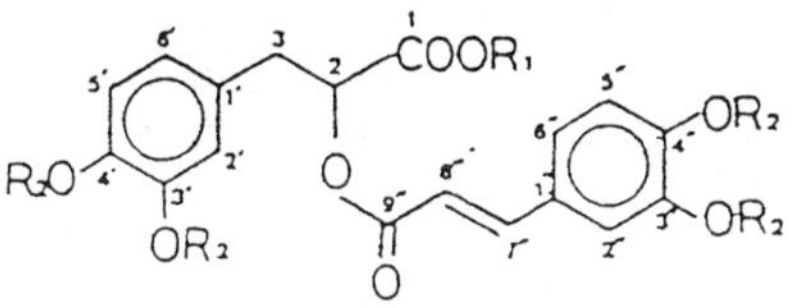

4a : $R_1 = R_2 = H$
4b : $R_1 = H$, $R_2 = COCH_3$
4c : $R_1 = CH_3$, $R_2 = COCH_3$

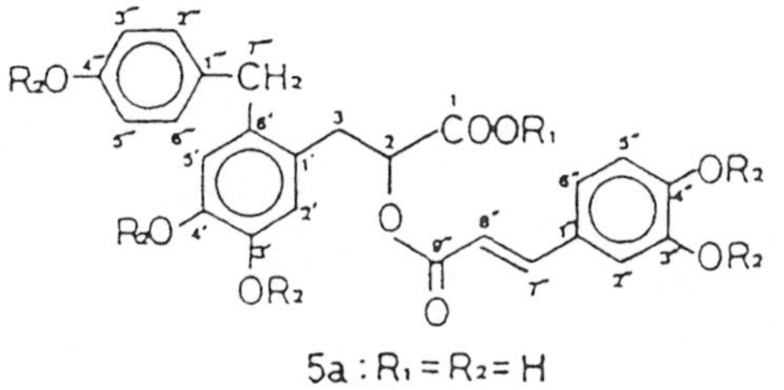

5a : $R_1 = R_2 = H$
5b : $R_1 = H$, $R_2 = COCH_3$
5c : $R_1 = CH_3$, $R_2 = COCH_3$

Figure 6.5 Structure of antioxidative compounds isolated from oregano: (1) 4-phenyl-β-D-glucopyranoside; (2) protocatechuic acid; (3) caffeic acid; (5a) 2-caffeoylpropinic acid. (From Refs. 40, 41.)

activities. The antioxidant principle of turmeric (the rhizome of *Curcuma longa* L.), which belongs to the same botanical family as ginger, was studied [47]. Some curcuminoids, including curcumin, 4-hydroxycinnmoyl (feruloyl) methane, and bis (4-hydroxycinnamoyl) methane were found to have antioxidant activities. Table 6.20 shows the 50% inhibitory concentration values (IC_{50}) of these and related compounds for air oxidation of linoleic acid. The IC_{50} values of these

TABLE 6.19 Antioxidative Actions of Mace, Nutmeg, and Their Petroleum Ether–Soluble and –Insoluble Fractions

			POV (mEq/kg)				
			100 hr	200 hr	300 hr	400 hr	500 hr
Control			2.7	7.3	13.5	35.6	356.5
Antioxidants	BHA	0.1%	—	—	1.1	3.2	12.2
	BHA	0.02%	—	3.1	14.0	16.0	26.1
	Tocopherol	0.1%	4.5	14.2	12.3	27.0	58.4
	Tocopherol	0.005%	4.2	6.4	7.2	31.2	276.5
Ground spices	Nutmeg	0.1%	2.2	6.1	11.5	27.2	205.6
	Mace	0.1%	1.7	3.9	5.7	11.2	13.7
Petroleum ether– soluble fractions	Nutmeg	0.1%	2.4	6.0	10.7	21.2	31.1
	Mace	0.1%	2.3	6.5	9.3	19.8	29.2
Petroleum ether– insoluble fractions	Nutmeg	0.1%	2.1	5.9	8.9	20.8	66.7
	Mace	0.1%	1.4	3.5	4.5	7.8	11.3

Antioxidative activities were evaluated by observing change of POV at 60°C.
Source: Ref. 32.

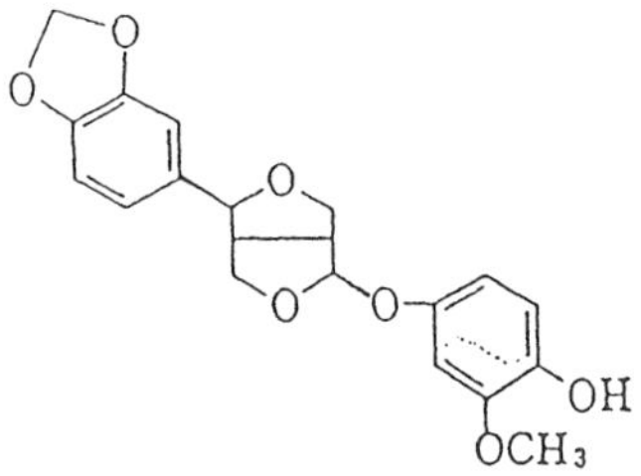

sesamolinol

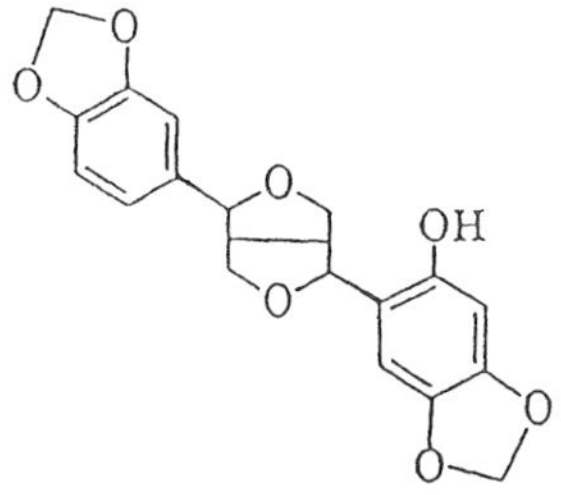

sesaminol

FIGURE 6.6 Structures of antioxidative compounds isolated from sesame seed. (From Refs. 43–45.)

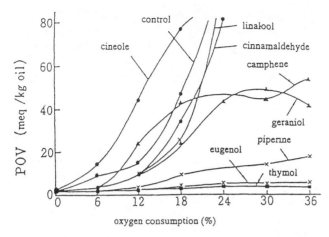

FIGURE 6.7 Spice compounds and their antioxidative effects on dehydrated pork during storage at 37°C. Each compound is added to pork with equivalent concentration of 0.2% of essential oil. (From Ref. 46.)

TABLE 6.20 IC_{50} of Antioxidative Compounds of Turmeric and Related Compounds on Air Oxidation of Linoleic Acid

Sample	50% inhibitory concentration (IC_{50})	
	TBAV (%)	POV (%)
Methanol ex.	1.22×10^{-2}	1.21×10^{-2}
Curcumin	1.83×10^{-2}	1.15×10^{-2}
4-Hydroxycinnamoyl(feruloyl)methane	1.88×10^{-2}	2.79×10^{-2}
Bis(4-Hydroxycinnamoyl)methane	2.80×10^{-2}	3.17×10^{-2}
Caffeic acid	5.63×10^{-3}	5.30×10^{-3}
Ferulic acid	8.95×10^{-3}	5.41×10^{-3}
Protocatechuic acid	1.85×10^{-2}	1.54×10^{-2}
Vanillic acid	2.01×10^{-2}	1.83×10^{-2}
BHA	3.37×10^{-3}	3.75×10^{-3}
BHT	1.92×10^{-3}	2.24×10^{-3}
dl-α-Tocopherol	1.95×10^{-1}	2.48×10^{-1}

Source: Ref. 47.

compounds were lower than that of α-tocopherol, meaning that they are more effective antioxidants. The antioxidant activities of these three compounds were found to be almost equal to that of protocatechuic acid, a strong antioxidant compound found in oregano, although they were less effective than BHA and BHT. Five biphenyl compounds have been isolated from thyme, two of which were confirmed to have antioxidant activity equal to BHT and one of which was as effective as α-tocopherol [48]. Capsaicin and Dihydrocapsaicin, the pungent principles of red pepper, were found to be responsible for its antioxidant activity. It has long been believed that black pepper has no antioxidant properties, but strong antioxidant activity was recently found in the acidic fraction of its extract and in the phenol amide compound.

D. Rosemary Extract

Rosemary extract has been commercialized as an antioxidant agent under the name of HSE: SP-100. The effectiveness of this rosemary agent was evaluated in various kinds of oils with the AOM test at 97.8°C and with the oven test at 60°C in comparison with commercially existing antioxidant agents: BHA and tocopherol [49]. As shown in Figure 6.8, the rosemary agent was found to retard the oxidation of lard to the same degree as BHA at a concentration of 200 ppm. Figure 6.9 shows the antioxidative activities of these agents on soybean oil. BHA and tocopherol did not exhibit antioxidant activity at a concentration of 200 ppm, whereas the rosemary extract did. This tendency was confirmed in the oven test, in which rosemary extract retarded oxidation longer than either BHA or tocopherol (Table 6.21). Rosemary extract was confirmed to be effective in other vegetable oils including rapeseed oil and corn oil, whereas BHA and tocopherol did not exhibit

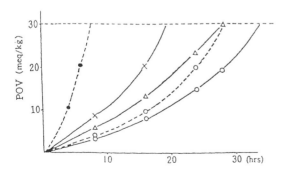

FIGURE 6.8 Antioxidative activity of rosemary agent, BHA, and tocopherol against lard. Effectiveness was evaluated with the AOM at 97.8°C. (—○—), Rosemary agent 400 ppm; (– –○– –), rosemary agent 200 ppm; (– –●– –), control; (—×—), tocopherol 200 ppm; (—△—), BHA 200 ppm. (From Ref. 49.)

Antimicrobial and Antioxidant Properties of Spices

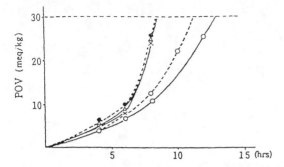

FIGURE 6.9 Antioxidative activity of rosemary agent, BHA, and tocopherol against soybean oil. Effectiveness was evaluated with the AOM at 97.8°C. (—○—), Rosemary agent 400 ppm;(– -○- –),rosemary agent 200 ppm; (- -●- -), control; (—×—), tocopherol 200 ppm; (—△—), BHA 200 ppm. (From Ref. 49.)

strong antioxidative activities when used in these vegetable oils (at 100 ppm). As shown in Figure 6.10, the antioxidant activity of tocopherol is not necessarily in proportion to the amount used, in fact the activity plateaus at higher concentrations, while the effectiveness of rosemary extract is enhanced proportionately with its concentration. Another advantage of rosemary extract is that a synergistic effect can be expected when it is used with tocopherol, while the antioxidant activity of BHA is not enhanced synergistically with added tocopherol.

TABLE 6.21 Antioxidative Activities of Rosemary Agent, BHA, and Tocopherol in Soybean Oil and Rapeseed Oil

Oil	Antioxidant	POV (mEq/kg)				
		0 day	3 days	5 days	7 days	9 days
Soybean oil	Control	2.1	4.8	15.5	27.5	44.7
	Tocopherol (200 ppm)	—	5.5	14.8	23.9	42.4
	BHA (200 ppm)	—	6.7	19.7	30.1	48.1
	Rosemary agent (100 ppm)	—	7.6	6.8	13.1	31.6
	Rosemary agent (200 ppm)	—	3.3	5.4	8.8	16.6
Rapeseed oil	Control	0.6	1.3	2.8	10.3	17.0
	Tocopherol (100 ppm)	1.9	1.2	3.4	9.5	18.2
	BHA (100 ppm)	2.6	1.2	2.1	9.1	15.8
	Rosemary agent (100 ppm)	0.6	1.0	1.7	5.1	11.9

Antioxidative activity was evaluated with the oven test at 60°C.
Source: Ref. 49.

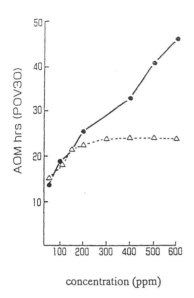

FIGURE 6.10 AOM hours (hours to reach peroxide value of 30 meq/kg) with several different concentrations of rosemary agent and tocopherol (80%) added to lard. Temperature of AOM was 97.8°C. (—●—), Rosemary agent; (- -△- -), tocopherol 80%.

III. EFFECTS OF SPICES ON INSECTS

Some spices have been reported to be effective when used to retard the growth of and/or disinfect against certain insects. A relatively large number of studies have reported on the activities of spices against, in particular, mites, dog roundworms, and parasitic worms.

A. Effects of Essential Oils of Spices Against Mites

Sixty to ninety percent of allergy sufferers are antigenic positive for mites. Of many types of mites, the antigenicities of *Dermatophagoides farinae* and *D. pteronyssinus* (House dust mites) are relatively strong. These mites, 0.2–0.4 mm in size, thrive at a temperature of 25°C and relative humidity (RH) of 75%. Yuri and Izumi [50], studying the effect of essential oils of various plants on *D. farinae*, found that essential oils of some spices were effective against this mite. They exposed mites to essential oils at concentrations of 80 μg/cm^2. The number of mites surviving after one day under optimum conditions (25°C, 75% RH) was counted, and the rate of mortality was calculated as follows:

$$\text{Mortality rate } (\%) = 100 \times \left(1 - \frac{\text{Number of mites in test dish}}{\text{Number of mites in control dish}}\right)$$

where test dishes contained essential oils and control dishes did not. The mortality rates of the essential oils of anise, cinnamon, bay leaves, clove, perilla, and pimento were found to exceed 50% (Table 6.22). Of the chemical compounds found in spice essential oils, cinnamaldehyde, eugenol, and isoeugenol exhibited high mortality rates (Table 6.23). The repellent/mortality rates of iso-compounds are in general lower than those of n-compounds. Carbon number and the presence of double bonds also seem to influence this effect.

Mansour et al. evaluated the essential oils from 14 spices of the Labiatae family for effectiveness against spider mites, a major pest of fruit trees and truck crops. All of the essential oils, including peppermint oil and lavender oil, were found to have repellent or toxic effects on this kind of mite [51].

B. Anthelmintic Effects of Spices

Tsuda et al. standardized the method for evaluation of anthelmintic activities [53]. Anthelmintic activity was expressed as the relative mobility (RM) value for larvae of dog roundworm (*Toxocara canis*), a typical nematode. The state of the larvae was scored according to the criteria described in Table 6.24, and the RM value was calculated from the following equations [52];

TABLE 6.22 Mortality Rate of Spice Essential Oils for *Dermatophagoides farinae*

Essential oil	Mortality rate (%)	Essential oil	Mortality rate (%)
Anise	56.5	Mace	0.5
Bay leaf	89.2	Nutmeg	—
Caraway	13.2	Thyme	27.5
Cardamom	4.7	Perilla	58.7
Clove	97.3	Pimento	58.2
Coriander	4.4	Rosemary	14.9
Cumin	14.9	Sage	12.0
Fennel	0.2	White pepper	0.1
Garlic	72.8	Capsicum oleoresin	6.7
Jasmine	9.2	Vanilla oleoresin	11.3

Concentration of each essential oil used in a filter paper was 80 μg/cm². Mortality rate was calculated by counting mites surviving one day after test started.
Source: Adapted from Ref. 50.

TABLE 6.23 Mortality Rate of Compounds Contained in Spice Essential Oils

Compound	Mortality rate (%)	Compound	Mortality rate (%)
Linalool	28.6	Iso-Eugenol	95.5
l-Carvone	13.2	Ethyleugenol	96.3
Benzaldehyde	3.1	Methyleugenol	37.6
α-Pinene	27.6	Safrol	2.5
Cinnamaldehyde	100	Vanillin	7.4
Eugenol	99.0		

Concentration of each compound was 80 μg/cm^2. Disinfecting rate was calculated by counting mites surviving one day after the test started.
Source: Adapted from Ref. 50.

$$MI = \frac{\Sigma\, nN_n}{\Sigma N_n}$$

where MI is the mobility index and N_n is the number of larvae with a score of n.

$$RM = \frac{MI_{sample} \times 100}{MI_{control}}$$

Each of 40 spices was extracted with water and methanol, respectively, and each extract was evaluated using the assay described above [53]. The RM values were calculated from the results obtained after 24 hours (Table 6.25). The RM values for both ethanol and water extracts of group A spices were zero, indicating that the spices in this group were very effective anthelmintic agents against dog roundworm. Anise, cinnamon, and clove were among the most effective anthelmintic spices. Group B and C spices exhibited anthelmintic activities in either their methanol extracts or their water extracts, and more than two-thirds of spices tested showed some inhibitory activity. It was also found that methanol extracts had in general stronger anthelmintic activity than water extracts. Other studies

TABLE 6.24 Criteria for Evaluating Effect on Larvae

State	Score (n)
Moving with whole body	3
Moving with only a part of body in the observation period	2
Immobile but not dead	1
Dead	0

Source: Ref. 52.

TABLE 6.25 Anthelmintic of Water and Methanol Extracts of Various Spices Against Dog Roundworm

Group	Water extract, 10 mg/ml	Methanol extract 10 mg/ml	Methanol extract 1 mg/ml	Spice
A	+	+	+	Anise, cinnamon, clove, mace, oregano, pepper, turmeric
B	−	+	+	Cardamom, cumin, ginger, nutmeg, Japanese pepper, thyme
C	−	+	−	Allspice, basil, caraway, coriander, dill, fennel, garlic, marjoram, rosemary, sage, tarragon
	+	−	−	
D	−	−	−	Red pepper, fenugreek, horseradish, mustard, paprika, parsley

+: RM value = 0; − RM value > 0. RM value was determined one day after assay was started.
Source: Adapted from Ref. 53.

revealed that the anthelmintic components of spices are not always contained in their essential oils. For example, the effective compounds in allspice were confirmed to be eugenol and methyleugenol, both in its essential oil, and tannin compounds, which remained in the residue after steam distillation.

More than 10,000 kinds of parasitic worms have been found in seafood. Most of these are not considered harmful to humans, but some might be harmful if seafood containing them is ingested in its fresh state. *Anisakis* larvae are sometimes parasitic on seafood that is used in sashimi, and many people have suffered from parasite diseases. *Anisakis* larvae are also parasitic on humans, living for several days in muscle or internal organs. People who have ingested these nematodes suffer within a few hours from vomiting and stomach pain.

Because of the continuous food chain in the ocean, almost any kind of seafood could contain *Anisakis* nematodes. The larvae of this parasite will die within a few hours at −20°C, therefore seafood frozen for this length of time should be relatively safe. But it is still necessary to be wary of cross-infection from tableware, utensils, or cutting boards, to which the larvae and the eggs can adhere.

Several studies have been conducted in Japan on spices used in seafood as a countermeasure to such parasitic worms. Yasuda, studying the inhibitory effects of several substances on the movement of *Anisakis* larvae [54], found cinnamalcohol and cinnamaldehyde to be effective. In another study, 12 kinds of foods and seasonings (e.g., perilla, wasabi, ginger, and garlic) were tested as to their ability to kill anisakids. The extract solutions contained 5% spice. The results for this study are shown in Table 6.26. Yasuda also studied individual components contained in perilla and ginger, which showed relatively high effectiveness. Peril-

TABLE 6.26 Anthelmintic Effects of Spices

Spice	Time required to stop movement of 10 anisakids	Effective component
Red perilla	3 hours	Perillaldehyde
Ginger	3 hours	Shogaol
Wasabi	5.5 hours	Allyl isothiocyanate
Garlic	11 hours	Diallyl disulfide
Salt water	2 weeks	

Concentration of each solution was 5%.

laldehyde was found to destroy anisakids at a concentration of 125 ppm, and shogaol was effective at half that concentration. However, gastric juices containing perillaldehyde were not as effective, meaning that it would be difficult to destroy *Anisakis* living on a human body only by taking a "normal" amount of perilla.

Consumer demand for organic vegetables has increased steadily, but insect control during cultivation is still difficult without using agricultural chemicals. The marigold plant has been used for growing radishes. The plant, including its flower (blooms from midsummer to late autumn), is plowed into the soil before radish seeds are sown. Using marigold in this way is said to suppress the breeding of the *Pratylenchus penetrans* nematode, which can be harmful to radish roots. Such cultivation methods in which effective plants are cultivated and plowed into soils during fallow periods are used mainly in Europe, where chemical pesticides are not as widely used.

REFERENCES

1. W. H. Martindale, *Perfum. Essent. Oil Rec.*, 1: 266 (1910).
2. A. R. Penfold and R. Grant, *Perfum. Essent. Oil Rec.*, 14: 437 (1923).
3. E. K. Rideal and A. Sciver, *Perfum. Essent. Oil Rec.*, 21: 341 (1930).
4. J. C. Maruzzella and M. B. Lichtenstein, *J. Am. Pharm. Assoc. Sci. Ed.*, 45: 378 (1956).
5. J. C. Maruzzella and M. B. Lichtenstein, *J. Am. Pharm. Assoc. Sci. Ed.*, 47: 250 (1958).
6. J. A. Morris, A. Khettry, and E. W. Seize, *JAOCS*, 56: 595 (1979).
7. L. A. Shelef, O. A. Naglik, and D. W. Bogen, *J. Food Sci.*, 56: 1042 (1980).
8. S. Ueda, H. Yamashita, M. Nakajima, and S. Kuwahara, *Nippon Shokuhin Kogyo Gakkaishi*, 29: 111 (1982).
9. A. Galli, L. Franzetti, and D. Briguglio, *Ind. Aliment.*, 24: 463 (1985).
10. S. Miyao, *Shokuhin Eisei Gakkaishi*, 16: 412 (1975).
11. K. Mori, H. Sawada, O. Nabetani, and S. Maruo, *Nippon Shokuhin Kogyo Gakkaishi*, 21: 286 (1974).

12. H. Dole and A. Knapp, Z. Hyg., 128: 696 (1948).
13. R. M. Julseth and R. H. Deibel, J. Milk Food Technol., 37: 414 (1974).
14. L. R. Beuchat, J. Food Sci., 41: 899 (1976).
15. C. N. Huhtanen, J. Food Prot., 43: 195 (1980).
16. Y. Kohchi, Food Chem., Nov. 34 (1995).
17. R. S. Farag, Z. Y. Daw, and S. H. Abo-Raya, J. Food Sci., 54: 74 (1989).
18. L. B. Bullerman, J. Food Sci., 39: 1163 (1974).
19. H. Hitokoto, S. Morozumi, T. Wauke, S. Sakai, and H. Kurata, Appl. Environ. Microb., 39: 818 (1980).
20. T. Katayama and I. Nagai, Bull. Jpn. Soc. Sci. Fish., 26: 29 (1960).
21. N. Kurita, M. Miyaji, R. Kurane, Y. Takahara, and K. Ichimura, Agric. Biol. Chem., 43: 2365 (1979).
22. T. Miyamoto, J. Antibact. Antifung. Agents, 14: 517 (1986).
23. K. Kanemaru and T. Miyamoto, Nippon Shokuhin Kogyo Gakkaishi, 37: 823 (1990).
24. K. Isshiki, K. Tokuoka, R. Mori, and S. Chiba, Biosci. Biotech. Biochem., 56: 1476 (1992).
25. C. J. Cavallito and J. H. Bailey, J. Am. Chem. Soc., 66: 1950 (1944).
26. A. Chain, Lancet, 241: 177 (1941).
27. K. Noda, S. Isozaki, and H. Taniguchi, Nippon Shokuhin Kogyo Gakkaishi, 32: 791 (1985).
28. D. L. Mavesty, U.S. patent 2, 124: 706 (1938).
29. C. W. DuBois and D. K. Tressler, Proc. Inst. Food Technol., : 202 (1943).
30. J. R. Chipault, G. R. Mizuno, J. M. Hawkins, and W. O. Lundberg, Food Res., 17: 46 (1952).
31. Y. Watanabe and Y. Ayano, Eiyo to Syokuryo, 27: 181 (1974).
32. Y. Saito, Y. Kimura, and T. Sakamoto, Eiyo to Syokuryo, 29: 505 (1976).
33. Y. Nitta, Chorikagaku, 10: 254 (1977).
34. C. H. Brieskorn, A. Fuchs, J. B. Bredenberg, J. D. McChesney, and E. Wenkert, J. Org. Chem., 29: 2293 (1964).
35. N. Nakatani and R. Inatani, Agric. Biol. Chem., 45: 2385 (1981).
36. R. Inatani, N. Nakatani, and H. Fuwa, Agric. Biol. Chem., 47: 521 (1983).
37. N. Nakatani and R. Inatani, Agric. Biol. Chem., 48: 2081 (1984).
38. C. M. Houlihan, C. T. Ho, and S. S. Chang, JAOCS, 61: 1036 (1984).
39. C. M. Houlihan, C. T. Ho, and S. S. Chang, JAOCS, 62: 96 (1985).
40. N. Nakatani and H. Kikuzaki, Agric. Biol. Chem., 51: 2727 (1987).
41. H. Kikuzaki and N. Nakatani, Agric. Biol. Chem., 53: 519 (1989).
42. W. M. Cort, Food Technol., 28: 60 (1974).
43. T. Osawa, M. Namiki, and Y. Fukuda, Agric. Biol. Chem., 49: 3351 (1985).
44. Y. Fukuda, T. Osawa, M. Nagata, M. Namiki, and T. Ozaki, Agric. Biol. Chem., 49: 301 (1985).
45. Y. Fukuda, M. Isobe, M. Nagata, T. Osawa, and M. Namiki, Heterocycles, 24: 923 (1986).
46. S. Fujio, A. Hiyoshi, T. Asari, and K. Suminoe, Nippon Shokuhin Kogyo Gakkaishi, 16: 241 (1969).
47. S. Toda, T. Miyase, H. Arichi, H. Tanizawa, and Y. Takino, Chem. Pharm. Bull., 33: 1175 (1985).

48. K. Miura and N. Nakatani, *Chem. Express*, 4: 237 (1989).
49. T. Kanda and T. Nakajima, *New Food Ind.*, 23: 36 (1981).
50. Y. Yuri and K. Izumi, *Aromatopia*, 3: 65 (1994).
51. F. Mansour, U. Ravid, and E. Putievsky, *Phytoparasitica*, 14: 137 (1986).
52. F. Kiuchi, N. Miyashita, Y. Tsuda, K. Kondo, and H. Yoshimura, *Chem. Pharm. Bull.*, 35: 2880 (1987).
53. Y. Tsuda and F. Kiuchi, *Koshinryo no Kinou to Seibun* (K. Iwai and N. Nakatani, eds.), Kouseikan, Tokyo, 1989, p. 189.
54. Yasuda, *Nihon Koshinryo Kenkyukai koen yoshishu, Vol. 12* (1995).

7

The Physiological Effects of Flavor/Aroma

I. THE EFFECT OF FLAVOR/AROMA ON THE HUMAN BRAIN

One of the most important properties of a spice is its flavor/aroma. The essential oils of spices, which are responsible for their aroma, are composed of many chemical compounds. It is estimated that around 400,000 kinds of chemical compounds make up natural flavors/aromas, more than 3000 of which have been identified so far. Although essential oils are widely used, only recently has research into how they are perceived been undertaken, including the effect of aroma on the brain [1–4].

A. Brain Waves and Olfaction

The fluctuations of electrical potential in the brain caused by stimulation of the nervous system are referred to as brain waves. The four types of brain waves, distinguished by frequency per second, are as follows: (a) alpha waves (8–13), indicating the subject is awake and resting; (b) beta waves (18–30), indicating intense nervous system activity; (c) delta waves ($< 3\frac{1}{2}$), indicating deep sleep or brain disorder; and (d) theta waves (4–7), usually found in adults under stress.

Flavor/aroma compounds in the air, which are volatile chemicals, travel through the nostrils and eventually into the lungs. These compounds are first dissolved in the mucus secreted by the Bowman's glands in the olfactory epithelium (see Figs. 7.1 and 7.2). Then, chemical receptors consisting of olfactory neurons and the connecting olfactory cells transmit electrical signals to the olfactory bulb in the brain.

B. The Effect of Aroma as Measured by Contingent Negative Variation

In a study in which the psychological effect of aroma was measured using brain waves, Torii [4] reported on the effect of aromatic essential oils on "the human

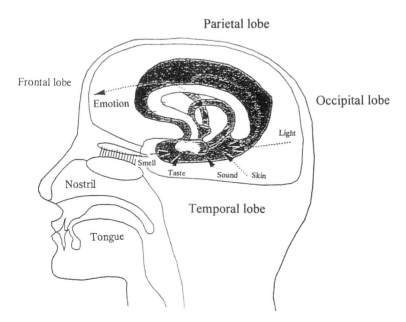

FIGURE 7.1 Olfaction and the limbic system.

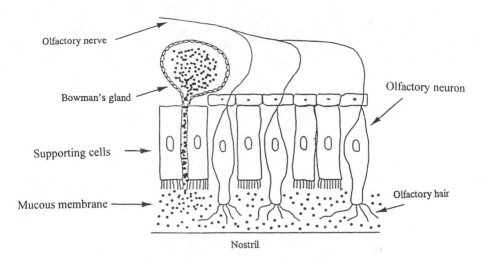

FIGURE 7.2 Structure of the olfactory epithelium.

living body." Psychological reaction was measured in this study by checking the brain waves in the frontal lobe of test subjects, who were asked to press a button when seeing a light turned on. The light was preceded several seconds by the sound of a buzzer, and the test subject, learning to anticipate the light, showed significant brain activity. The measurement of these anticipatory brain waves has been referred to as contingent negative variation (CNV). The method of measuring brain waves by CNV was reported by Walter et al. [5].

Figure 7.3A shows the change in the test subject's brain waves when the buzzer was perceived: large waves appear and quickly disappear again [6]. Figure 7.3B shows the change recorded when the light was turned on: again, large waves were temporarily recorded. Figure 7.3C shows the brain waves recorded when the buzzer and light occurred consecutively. In this case, the waves observed correspond to those in both Figure 7.3A and 7.3B. Figure 7.3D shows the wave change when the test subject was asked to press a button when the light was turned on. The recorded brain waves move up into the negative area while the subject anticipates pressing the button soon after hearing the buzzer.

It is known that the CNV increases when the brain is stimulated, but it decreases when the brain is in a calm state. In fact, the CNV increases if one drinks a stimulant like coffee (Fig. 7.4).

The effect of aroma on the brain can also be tested by observing the CNV pattern. One experiment [4] compared the essential oils of jasmine and lavender (*Lavandula officinalis* Mill) with a control, in which no essential oil was contained. The results of this experiment showed that the CNV for jasmine was

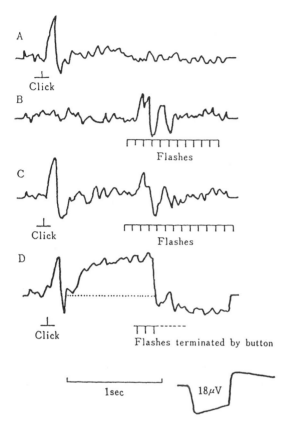

FIGURE 7.3 Example of CNV pattern. (From Ref. 6.)

greater than that for the control, whereas the CNV for lavender was less than that for the control (Fig. 7.5). This confirms the long-believed characteristics of these two spices: jasmine as a stimulant and lavender as a sedative. Only using CNV could these qualities be scientifically confirmed.

Torii examined the effects of the essential oils of a variety of medicinal herbs and spices on CNV [3]. As shown in Figure 7.6, the CNV value for a case of no essential oil was set at 100. The smaller the CNV value the higher the sedative effect of that essential oil. Or, the higher the CNV value, the more stimulating the essential oil. Sedative effects were reported for the essential oils of lemon, geranium, and chamomile, which are herbs, whereas the essential oils of the spices clove, basil, and peppermint were found to be stimulants. Caraway, marjoram, rosemary, and spearmint were also confirmed to possess sedative effects. It is possible to generalize that a number of spices have stimulative effects, while a number of herbs act as sedatives.

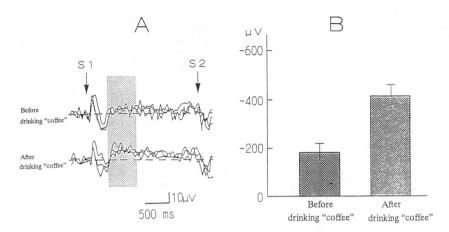

FIGURE 7.4 Effect of coffee on CNV. (From Ref. 3.)

C. Making Use of the Stimulant Effects of Essential Oils of Spices

One study examined the theory that peppermint aroma in combination with the sound of a buzzer prevent drivers from falling asleep at the wheel [7]. Automobile companies are conducting research into so-called "doze-detecting systems." Such systems can determine the degree of wakefulness by observing eye movement or eyelid position.

One test was conducted in which test subjects on the verge of sleep awakened by a buzzer. It was determined that the degree of wakefulness decreased as the test subjects became accustomed to the sound of the buzzer. Then the aromas of peppermint, jasmine, lavender, and lemon were blown at the test subjects on the verge of sleep. The aroma of peppermint was found to have the strongest stimulating effect, and jasmine was also confirmed to be effective (with individual differences). However, there was no significant difference between the results for lemon, lavender, or no aroma at all. It was also observed how long the test subject remained awake after each treatment. Test subjects were found to stay awake for more than 10 minutes after hearing the buzzer and smelling the peppermint, whereas they stayed awake only a minute after exposure to either the buzzer or the peppermint aroma alone. It showed that peppermint is more effective when it is used after people's waking up completely than when it is used before waking up.

The brain waves of six male test subjects were observed when their states of sleepiness were challenged using different means. The total beginning electrical potential of twelve electrodes was set to be 100, and it decreased to 74.4 when the test subjects started to feel sleepy. When the test subject was given chewing gum, the electrical potential jumped to 150, with coffee it reached 139, and when

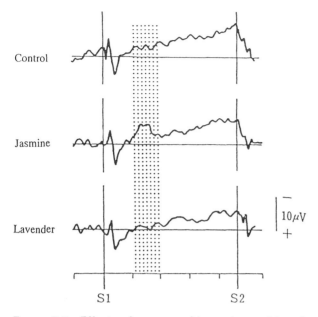

FIGURE 7.5 Effects of aromas of lavender and jasmine on CNV pattern. S1: First stimulus (buzzer); S1: second stimulus (flush). (From Ref. 4.)

the subject was asked to sing, it went up to 120. But after 10 minutes, the total electrical potential for chewing gum was still around 124, while for coffee it had decreased to normal, and for singing it actually dropped to 81. Therefore, no means other than chewing gum produced a continuous waking effect. Among different kinds of chewing gum, a menthol-flavored gum, in which peppermint and spearmint were the major ingredients, was the most effective. The effectiveness of menthol-flavored gum was in general twice that of coffee-flavored gum and four times that of chewing gum containing chlorophyll.

D. The Effect of Flavor/Aroma on Memory

The physical change that occurs in the brain when something is learned is called a memory scar, or engram. Memories can be triggered, sometimes involuntarily, by various physiological stimuli, including aroma.

Sakurai et al. [8] studied the effect of aroma on memory. One hundred test subjects were shown single, unrelated words while being exposed to one of several kinds of aroma as well as no aroma. When retested the following day, it was found that test subjects exposed to the aromas of rosemary or chocolate retained more than subjects exposed to no aroma. In experiments with mice conducted by Saito et al. [9], saffron was also found to improve memory scarring.

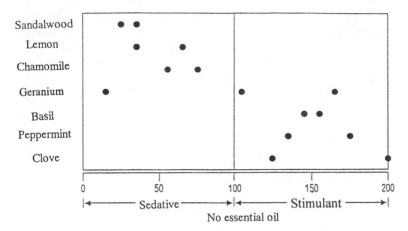

FIGURE 7.6 Effect of essential oils on CNV. (From Ref. 3.)

II. THE EFFECT OF SPICE FLAVOR/AROMA ON IMMUNE FUNCTION AND THE NERVOUS SYSTEM

A. Spice Flavor/Aroma and Anti-Stress Function

The concept of aromatherapy, a term said to have originated with the French pathologist Gatterfosse in the 1930s, has attracted attention lately. The essential oils of various spices have been proven to permeate the skin and have subsequent physiological effects.

Subjects suffering from physical or mental stress exhibit various physiological symptoms, including changes in blood pressure and hormone imbalances. For example, high blood pressure, headaches, and gastrointestinal disorders are known to be closely related to stress.

It was Selye who first conceptualized "stress" as having physiological effects [10]. In studies conducted with rats, he found the following:

1. At first, rats exposed to stress have a decreased resistance to trauma—a period of reaction.
2. As stress continued, various systems (e.g., nervous, immune) provided a period of resistance.
3. If stress continued further rats ceased to resist, falling ill and sometimes dying.

It is well known that stress affects the immune reaction, and studies of the relationship between aroma and immune function have been conducted. In studies with mice [3], lymph production was decreased when animals were restrained for

15 minutes, but immune function was recovered when animals were exposed to the aroma of α-pinene. In another experiment, mice injected with foreign cells (red corpuscles from sheep) managed to react by producing a certain number of antibody-producing cells. The number of antibody-producing cells was decreased by half when mice were restrained (subjected to stress). But if mice were exposed to α-pinene after a decline in immune function, that function returned to normal. α-Pinene was not, however, effective for the "normal" mouse not exposed to any stress (Fig. 7.7).

B. The Effect of Spice Flavor/Aroma on the Autonomic Nervous System

Aroma has an effect on the autonomic nervous system as well as on brain cells. This includes the sympathetic nervous system, which tends to act during stress and excitement. A study by Torii et al. examined the relationship between CNV and skin potential level (SPL), which is a measure of the electrical potential of the skin and an index of activity of the sympathetic nerves. Figure 7.8 shows that electrical potential decreased when subjects relaxed, but that when subjects smelled the aroma of jasmine, the electrical potential did not decrease. This electrical potential started to decline if the aroma was eliminated from the atmosphere. On the other hand, when test subjects smelled the aroma of chamomile, the electrical potential dropped, indicating a decrease in sympathetic nervous system activity.

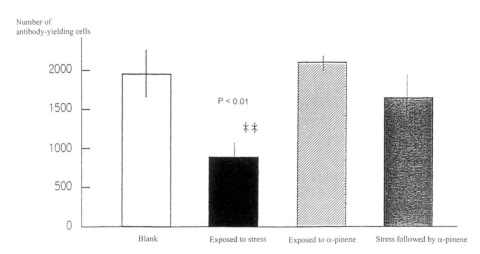

FIGURE 7.7 Effect of stress and/or α-pinene on production of antibody-producing cells. (From Ref. 4.)

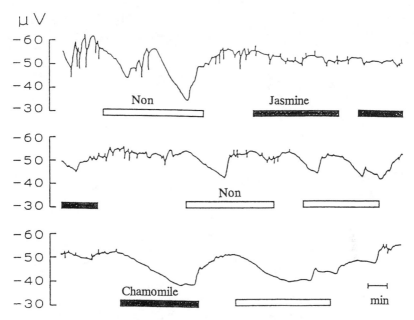

FIGURE 7.8 Effect of aroma on skin potential level. (From Ref. 3.)

C. Convulsion-Inhibiting Action of Essential Oils

Essential oils, which have an affinity with lipids, are easily absorbed from the skin as well as from the throat, nose, and mucous membrane of the stomach and intestines, then into blood, and work directly into brain cells. One study confirmed that α-pinene was detected in expired air 20 minutes after it was inhaled [12]. When lavender oil was rubbed into skin, linalool and linalool acetate, both of which are the major compounds of lavender oil, were detected in the blood within 20–30 minutes; it disappeared from blood in 90 minutes (Fig. 7.9). The compounds that permeate the skin are taken into the capillary vessels at the skin surface and then conveyed throughout the body. Since the flow of blood in the skin is relatively slow in comparison with that in muscle, the degree of absorption is controlled. Massaging the essential oil into the skin should enhance uptake by the blood stream. The fact that the central nervous system, which is rich in lipid, has a high affinity for essential oil should encourage aromatic compounds to pass from the blood to the brain cells.

Aromatic compounds accumulated in brain cells would affect the permeability of ions. The levels of calcium and other ions are important in the physiology of convulsions, and these can be affected by levels of aromatic compounds in the brain. Even single essential oils can contain compounds that work antago

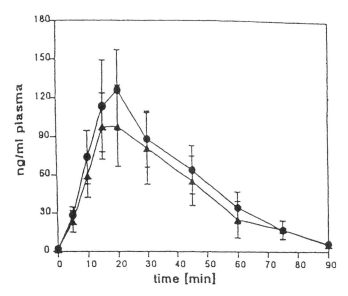

FIGURE 7.9 Percutaneous absorption of linalool and linalool acetate, which are found in essential oil of lavender. ●: Linalool; ▲: linalool acetate. (From Ref. 12.)

nistically. For example, the major compounds found in rosemary oil are α-pinene and 1,8-cineole. 1,8-Cineole works to break down acetylcholine, which has the effect of inhibiting convulsion of the small intestine, whereas α-pinene would cause a convulsion. How much convulsions were inhibited would depend upon the concentration of 1,8-cineole contained in the rosemary oil.

The inhibitory effect of a compound on a convulsion is observed in the smooth muscle, which covers the stomach, small intestine, bronchi, and blood vessels. Contraction of the smooth muscles is promoted by increasing the amount of calcium ions in their amount. But if the inflow of calcium ions is inhibited, the contraction of the smooth muscle is suppressed. Menthol, a major compound of peppermint, works as a modulator of the calcium channel systems, therefore suppressing convulsions. It is for this reason that peppermint is effective against digestive problems, diarrhea, and headaches.

III. THE EFFECT OF FLAVOR/AROMA COMPOUNDS ON HUMAN SEX HORMONES

The apocrine gland of the underarm, which becomes active at puberty, is among the glands that secrete hormones. The secretion discharged from this gland is

The Physiological Effects of Flavor/Aroma

initially odorless, but is acquires an odor due to interaction with microbes. The main ingredients of this odor are androsterone or androstenol, which are derivatives of the male sex hormone testosterone.

Okazaki [11] studied the effect of androsterone and androstenol on the arousal levels of both males and females. In this study, the CNV for each odor was measured from the frontal lobe to observe variations in arousal levels between a control (no odor) and odor (Fig. 7.10). It was found that androsterone had a sedative effect on males and a stimulant effect on females, whereas androstenol had the opposite effects.

Both estrogen and progesterone are important female sex hormones produced by the ovaries after the female has reached puberty. Most female sex hormones have chemical structures based on cholesterol (Fig. 7.11), which allow them to bind with special receptors. Some of the components contained in essential oils are known to have chemical structures similar to those of some hormones and to work like hormones. Anetol, which is contained in fennel seed and anise seed, is a methylester of estrone and has a powerful estrogenlike effect. It was confirmed almost 60 years ago that the essential oils of both anise and fennel had a stimulating effect on the secretion of milk by the female. It is thought to be possible to utilize the essential oils with such estrogenlike activities in treating hormone imbalances.

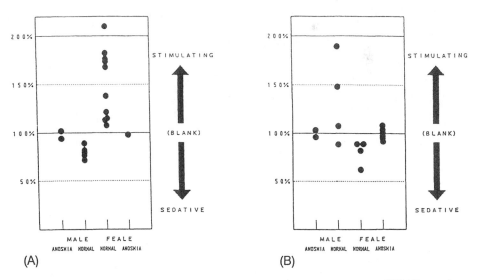

FIGURE 7.10 Effects of (A) androstenol and (B) androsterone on CNV (frontal region). (From Ref. 11.)

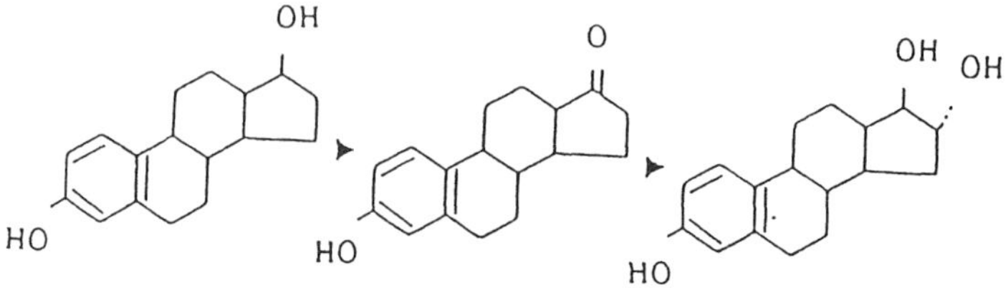

Estradiol Estrone Estriol

(Ovarian follicle hormone)

Estrogen

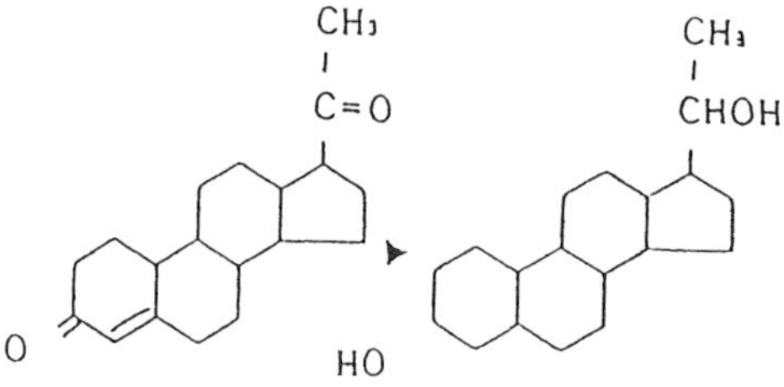

Progesterone Pregnanediol

(Corpus luteum hormone)

Figure 7.11 Chemical structure of female sex hormones.

REFERENCES

1. H. Fukuda, Y. Torii, H. Kanamoto, T. Miyauchi, and Y. Hamanabe, *The outline of "Aji to Nioi no symposium"*, Vol. 19, 65 (1985).
2. S. Ogata, T. Nishimura, M. Indo, M. Kawasaki, and Y. Torii, *The outline of "Aji to Nioi no symposium"*, Vol. 20, 149 (1986).
3. Y. Torii, Kaori no Hito ni Ataeru Eikyo, *Kateikagaku Kenkyu*, Lion Home Science Research, Tokyo, 1993, p. 201.
4. Y. Torii, *Kaori no Nazo*, Fragrance Journal, Tokyo, 1994.
5. W. G. Walter, P. Cooper, V. J. Aldridge, W. C. McCallum, and A. L. Winter, *Nature*, 203: 380 (1964).
6. T. Takiguchi, *Food Chem.*, 10(12): 78 (1994).
7. M. Hiramatsu, *Mainichi Shibun*, May 6: (1996).
8. H. Sakurai, S. Koiwa, M. Ito, and T. Tsurumi, *Saimin to Kagaku*, 9: 27 (1994).
9. H. Saito and N. Nishio, *Quark*, (3): 32 (1996).
10. H. Selye, *Nature*, 138: 32 (1936).
11. Y. Okazaki, *Fragrance J.*, 20(10): 80 (1992).
12. W. Jager, G. Buchbauer, L. Jirovetz, and M. Fritzer, *J. Soc. Cosmet. Chem.*, 43: 49 (1992).

Index

Aging, 63
Ajoene, 155
Alizarin, 22
Alkyl cysteinsulfide, 10
Alkyl disulfide, 10
Allicin, 10, 155, 156
Alliin, 10, 11, 155
Allium plant, 10, 12
Allspice, 2, 5, 6, 62, 70, 78, 92, 95, 99, 102, 103, 104, 105, 117, 119, 167
    antimicrobial effect, 167
    bitterness, 70
    deodorizing effect, 78

Allyldisulfide, 188
Allyl isothiocyanate, 9, 142, 146, 177, 179, 180
    antimicrobial effect, 177, 179, 180
    catecholamine secretion, 142, 146
Allynase, 10
Anethole, 75
Anise, 2, 5, 6, 74, 75, 78, 90, 92, 95, 99, 102, 103, 105, 122, 125, 165, 195, 196, 197
    anthelmintic effect, 196, 197
    antimicrobial effect, 164
    deodorizing effect, 78
    effect against mites, 195

Anisole deodorizing effect, 81
Annatto, 22, 25
Anthelmintic effect, 195–198
Anthocyanin, 13, 21
Anti-HIV function, 158–161
Antioxidant effect, 5, 77
Aspergillus, 164, 165, 172, 173, 174, 175
Astaxanthin, 17, 18

Bacillus subtilis, 164–166
Basil, 2, 5, 6, 75, 78, 92, 95, 99, 102, 103, 104, 105, 111, 113, 122, 123, 204
   deodorizing effect, 78
Bay leaves, 2, 5, 6, 70, 80, 82, 92, 93, 99, 100, 102, 103, 104, 105, 111, 126, 129, 195
   bitterness, 70
   deodorizing effect, 80, 82
   effect against mites, 195
Benzaldehyde, deodorizing effect, 81
β-carotene, 4, 13, 15, 16, 17, 18
Betanine, 23–24
BHA, 27, 182, 183, 185, 186, 187, 190, 191, 192, 193
BHT, 27, 182, 187, 190, 191, 192
Bitterness compounds, 70
Bixin, 22
Black pepper (*see* pepper)
Blending effect, 63–63
Borneol, 70
   antimicrobial effect, 175, 176
Bouquet garni, 64
Brown adipose tissue, 152–153
Butylcrotonyl isothiocyanate sulfide, 9

Cacao, 23, 25
Caffeic acid, 186, 189
Cantaxanthin, 17
Capsaicin, 9, 72–73, 142–155, 157
   absorption, 142, 143
   effect on lipid metabolism, 150, 153–154

[Capsaicin]
   effect on perirenal adipose tissue, 150–152
   effect on rectal temperature, 157
   effect on serum triglyceride, 150, 153
   effect on the preference for salty taste, 72–73
   epinephrine secretion, 144, 146
Capsantin, 13, 15, 19–20
Capsorbin, 13, 15
Caramel, 25
Caraway, 2, 5, 6, 67, 68, 69, 70, 78, 102, 103, 107, 130, 132, 170–172, 204
   antimicrobial effect, 164
   bitterness, 70
   deodorizing effect, 67, 68, 69, 78
   inhibitory effect for aflatoxin production, 170–172
Cardamom, 2, 5, 6, 58, 67, 70, 74, 78, 102, 103, 108, 112, 113, 126, 127
   bitterness, 70
   deodorizing effect, 67, 78
   effect on the sweet sensation, 74
Carnosol, 186, 187
Carotenoids, 13, 17
Carotene contents, 17
Carrot, 17
Carsonol, 77
Carthamin, 25
Carvacrol, 70
   antimicrobial effect, 175, 176, 179
Carvone, 70
Cassia, 79
Catecholamine secretion, 142, 145
Celery, 1, 6, 67, 68, 69, 78, 92, 93, 95, 99, 102, 103, 104, 105, 108, 111, 115, 118
   deodorizing effect, 67, 68, 69, 78
Chamomile, 204, 208
Chemical deodorizing, 75–76
Chili powder, 62, 63, 64
Chinese five spice, 63, 64
Chlorophyll, 13, 15, 25, 77, 79
   deodorizing effect, 77, 79

Index

Cineol, 188, 191, 210
  antioxidant effect, 188, 191
  convulsion-inhibiting effect, 210
Cinnamaldehyde, 75, 175, 177. 195, 196
  antimicrobial effect, 175, 177
  effect against mites, 195, 196
Cinnamon, 1, 4, 5, 6, 62, 66, 67, 70, 74, 75, 80, 90, 95, 90, 95, 99, 102, 103, 104, 105, 110, 112, 115, 164–167, 173, 195, 196
  anthelmintic effect, 196
  antimicrobial effect, 164–167, 173
  bitterness, 70
  deodorizing effect, 67, 80
  effect against mites, 195
  effect on the sweet sensation, 74
  texture improvement, 4
Classification of spice, 2
Clostridium botulinum, 165, 167, 169
Clove, 2, 5, 6, 61, 62, 67, 68, 69, 70, 78–79, 74, 90, 92, 99, 100, 102, 103, 104, 105, 111, 126, 129, 164–167, 170–172, 196, 197, 204
  anthelmintic effect, 196, 197
  antimicrobial effect, 164–167
  bitterness, 70
  deodorizing effect, 67, 68, 69, 78–79
  inhibitory effect for aflatoxin production, 170–172
Cochineal, 25
Condiments, 3
Contingent negative variation, 202–204
Cooling effect, 149
Coriander, 1, 5, 6, 67, 68, 69, 70, 90, 92, 95, 97, 100, 102, 112, 113, 130, 134
  antimicrobial effect, 164
  bitterness, 70
  deodorizing effect, 67, 68, 69, 70
Crocetin, 15, 20–21
Crocin, 15, 20
Cryptoxanthin, 15, 17
Cumin, 1, 5, 6, 70, 78, 102, 103, 105, 107, 112, 117, 121, 170–172
  bitterness, 70

[Cumin]
  deodorizing effect, 78
  inhibitory effect for aflatoxin production, 170–172
Cumin aldehyde, 70
Curcumin, 15, 19
Curry powder, 63, 64, 82
  deodorizing effect, 82

Definition of spice, 1–3
Diallyl sulfide, 9, 179
  antimicrobial effect, 179
Diallyldisulfide, 9, 11, 142, 146, 198
  anthelmintic effect, 198
  catecholamin secretion, 142, 146
Diallylthiosulfonate, 11
Diallyltrisulfide, 11
Dihydrocapsaicin, 9, 142
  absorption, 142
Dill, 2, 5, 6, 70, 78, 90, 92, 95, 99, 102, 103, 104, 105, 112, 117, 123, 164
  antimicrobial effect, 164
  bitterness, 70
  deodorizing effect, 78
Dimethyl sulfide, 8
Double layer coating spice, 56, 58
Dry soluble spice, 56

Eicosapentaenoic acid, 156
Ephedrine, 152
Epirosmanol, 186
Escherichia coli, 164–166, 178, 179, 180
Eugenol, 70, 81, 172, 174, 188, 191, 195, 196
  antimicrobial effect, 172, 174
  antioxidant effect, 188, 191
  effect against mites, 195, 196
European grape, 23

Fennel, 2, 5, 6, 74, 75, 78, 90, 92, 95, 99, 102, 105, 111, 122, 125
  deodorizing effect, 78

Fenugreek, 2, 78, 95, 99, 102, 107, 111, 126, 127
  deodorizing effect, 78
Flavonoid, 13

Garam masala, 64
Gardenia, 20, 25
Garlic, 2, 5, 6, 9, 10, 12, 67, 68, 69, 74, 82, 92, 99, 100, 102, 103, 104, 105, 112, 126, 128, 155–158, 165, 179–181
  antimicrobial effect, 165, 179–181
  deodorizing effect, 67, 68, 69, 82
  physiological effects, 155–158
  pungency, 9, 10, 12, 14
Ginger, 1, 5, 6, 9, 13, 14, 15, 66, 67, 78, 79, 80, 82, 90, 92, 95, 97, 100, 102, 103, 108, 112, 135, 137, 180, 183, 184, 185
  antimicrobial effect, 180
  antioxidant effect, 183, 184, 185
  color, 15
  deodorizing effect, 67, 78, 79, 80, 82
  pungency, 9, 13, 14
Glutamic oxaloacetic transaminase, 155
Glutamic pyruvic transaminase, 155
Glycogen, 142

Hibiscus, 21
Horseradish, 2, 5, 9, 10, 12, 13, 14, 82, 92, 99, 100, 102, 103, 113, 135, 137
  deodorizing effect, 82
  pungency, 9, 12, 13, 14
Hot sensation, 10
Hydroxybenzaldehyde deodorizing effect, 81
Hydroxybenzyl isothiocyanate, 9

Indigo, 23
Interscapular brown adipose tissue, 152, 156, 158
Isoborneol, antimicrobial effect, 175, 176
Isoeugenol, 195, 196

Isothiocyanate compound, 77, 158, 159
  deodorizing effect, 77
  effect on interscapular brown adipose tissue, 158
  effect on rectal temperature, 159

Japanese pepper, 2, 5, 9, 13, 14, 78, 82, 92, 97, 100, 102, 107, 111, 112, 135, 138
  deodorizing effect, 78, 82
  pungency, 9, 13, 14
Japanese seven spices (*see* shichimi)
Jasmine, 203, 204, 205, 206, 208

Lavender, 92, 203, 204, 205, 206, 209
  convulsion-inhibiting effect, 209
Leek, 2, 10, 17, 13
  pungency, 13
Linalool, 70, 188, 209
  antioxidant effect, 188, 191
  convulsion-inhibiting effect, 209
Linalool acetate, 209
Litmus moss, 23
Locked in spice, 56
Lutein, 14–15
Lycopene, 17

Mace, 5, 6, 70, 90, 92, 95, 99, 102, 112, 122, 124, 165–169, 183, 184, 185, 186–187, 190
  antimicrobial effect, 165–169
  antioxidant effect, 183, 184, 185, 186–187, 190
  bitterness, 70
  effect on slime formation, 165, 169
Madder, 22, 25
Marjoram, 2, 5, 6, 70, 78, 90, 92, 95, 102, 104, 111, 122, 165–167, 183, 185, 204
  antimicrobial effect, 165–167
  antioxidant effect, 183, 185
  bitterness, 70
  deodorizing effect, 78

Index

Masking action, 8
Masking effect, 5, 75–76
Menthol, 210
Mercaptan, 13, 77, 81–82, 83
Methyl allyltrisulfide, 155, 156, 160
   effect on thrombus formation, 155–156, 160
Methyl chavicol, 75
Methyl mercaptan (*see* mercaptan)
Methylene chloride extraction, 59–60
Mint, 1, 5, 78, 90, 92, 95, 99, 102, 103, 104, 105, 113, 117, 120
   deodorizing effect, 78
Monascus, 25
Mustard, 1, 5, 9, 12, 13, 14, 15, 17, 75, 77, 82, 92, 99, 100, 102, 104, 105, 135, 136, 177–179
   antimicrobial effect, 177–179
   color, 15, 17
   deodorizing effect, 75, 77, 82
   pungency, 9, 12, 13, 14
Myristphenone, 187

Neoxanthin, 15
Norbixin, 22
Norepinephrine, 147
Nutmeg, 2, 5, 6, 67, 70, 80, 90, 92, 95, 99, 100, 102, 103, 104, 105, 112, 126, 130, 161, 167, 183, 184, 185, 186–187, 190
   antimicrobial effect, 167
   antioxidant effect, 183, 184, 185, 186–187, 190
   bitterness, 70
   deodorizing effect, 67, 80
   physiological effect, 161

Offset deodorizing, 75–76
Onion, 2, 5, 9, 10, 13, 14, 21, 25, 61, 67, 74, 79, 82, 92, 93, 99, 100, 102, 103, 104, 105, 111, 112, 126, 128
   deodorizing effect, 67, 82
   pungency, 9, 13, 14

Oregano, 2, 5, 6, 70, 74, 78, 92, 99, 100, 102, 103, 104, 105, 112, 112, 130, 133, 165, 183, 185, 186
   antimicrobial effect, 165
   antioxidant effect, 183, 185, 186
   bitterness, 70
   deodorizing effect, 78

Paprika, 2, 4, 5, 13, 15, 16, 17, 18, 19–20, 25, 27, 92, 95, 99, 101, 102, 103, 105, 135, 138
   color, 4, 15, 16, 17, 25
   color stability, 27
Parsley, 2, 5, 15, 17, 18, 92, 93, 95, 99, 102, 103, 104, 108, 112, 113, 115, 118
Pepper, 2, 5, 6, 58, 61, 66, 71–72, 92, 95, 99, 100, 102, 103, 104, 105, 130, 135, 164, 167, 168,
   antimicrobial effect, 164, 167, 168
   deodorizing effect, 67, 78, 82
   flavor, 58
   pungency, 9, 13, 14
Peppermint, 7, 90, 204, 205, 206, 209
   convulsion-inhibiting effect, 209
   microscopic picture, 7
Peranin, 20
Perilla, 17, 25, 78
   deodorizing effect, 78
Perillaldehyde, 175, 177, 197
   anthelmintic effect, 197
   antimicrobial effect, 175, 177
Phenol deodorizing effect, 80, 81
4-Phenyl-β-D-glucopyranoside, 186, 189
Physical deodorizing, 75–76
Pickling spice, 64
α-Pinene, 208, 209, 210
   convulsion-inhibiting action, 209, 210
   effect on stress, 208
Piperine, 9, 142, 146
   absorption, 142
   catecholamine secretion, 142, 146
Piperonal, 81
Polyphenol compound, 70

Protocatechuic acid, 186, 189
Purprin, 22–23

Quatre épices, 64
Quercetin, 21

Radish
  deodorizing effect, 82
  pungency, 9, 13, 14, 82
Red beet, 23, 25
Red pepper, 2, 5, 8, 9, 13, 14, 15, 17, 18, 19–20, 92, 95, 99, 100, 102, 103, 105, 112, 135, 136, 180
  antimicrobial effect, 180
  color, 15, 17, 18
  pungency 9, 13, 14
Red perilla, 20
Rosmanol, 77, 186, 187
Rosmaridiphenol, 186, 188
Rosmariquinone, 186, 188
Rosemary, 2, 5, 70, 77, 79, 92, 95, 99, 100, 102, 103, 104, 105, 111, 112, 130, 132, 182–188, 192–195, 204, 206, 209–210
  antimicrobial effect, 164
  antioxidant effect, 77, 182–188, 192–195
  bitterness, 70
  convulsion-inhibiting effect, 209–210
  deodorizing effect, 79
  effect on memory, 206
Rutin, 13, 15, 17, 18

Sedative effect of spice, 204
Safflower, 25, 26
Saffron, 2, 4, 5, 15, 17, 20, 56, 92, 95, 99, 101, 102, 104, 107, 111, 139, 140
  color, 4
  effect on memory, 206
Safrole, 26
Sage, 2, 5, 6, 67, 68, 69, 70, 74, 77, 78–79, 80, 82, 90, 92, 95, 99,

[Sage]
  100, 102, 103, 104, 105, 112, 130, 133, 164, 182–185
  antimicrobial effect, 164
  antioxidant effect, 77, 182–185
  bitterness, 70
  deodorizing effect, 67, 68, 69, 78–79, 80, 82
Salicylic aldehyde, 175, 176, 179
Sanshool, 9
Scordinin, 155, 156–157, 180
  antimicrobial effect, 180
Sensational deodorizing, 75–76
Serum glucose, 142, 144
Sesame, 1, 8, 92, 95, 99, 102, 111, 122, 125, 188, 190
  antioxidant effect, 188, 190
Sesamiol, 188, 190
Sesamolinol, 188, 190
Sharp sensation, 10
Shichimi, 8, 63, 64
Shogaol, 9
Silver salmon, 17
Sisonin, 20
Slime formation, 165, 168
Sodium copper chlorophyllin (*see* chlorophyll)
Soirulina, 25
Spearmint, 204
Spice emulsion, 55, 56
Spice seasonings, 3
Spinach, 17
Stability of color, 26
Staphylococcus aureus, 164–167, 179–180
Star anise, 2, 6, 74, 75, 78, 92, 95, 99, 102, 103, 111, 122, 123
  deodorizing effect, 78
Steam distillation, 59–60
Streptomyces, 164
Substance P, 142
Sugar content of spice, 74
Sulfide compounds, 13
Supercritical fluid extraction, 59–60
Suppressive effect, 63–67
Sweet flavor compounds in spice, 74–75

Index

Sympathetic nerve, 148
Synergistic effect, 63–67

"Tade", 9, 14
Tadenal, 9
Tarragon, 1, 6, 92, 95, 99, 102, 103, 104, 105, 113, 117, 121
Terpineol, 70
Thyme, 1, 5, 6, 67, 68, 69, 70, 78–79, 90, 92, 95, 99, 100, 102, 103, 104, 105, 111, 112, 130, 131, 165–167, 169, 170, 171, 172, 183, 185
  antimicrobial effect, 165–167, 169, 170
  antioxidant effect, 183, 185
  bitterness, 70
  deodorizing effect, 67, 68, 69, 78–79
  inhibitory effect for aflatoxin production, 170–172
Thymol, 70, 175, 176, 188, 191
  antimicrobial effect, 175, 176
  antioxidant effect, 188, 191
Tocopherol, 27, 187, 188, 192, 193
Trimethylamine, 8, 78–80
Turmeric, 1, 5, 15, 19, 25, 78, 92, 95, 95, 99, 101, 102, 103, 108, 112, 135, 139

[Trimethylamine]
  antioxidant effect, 189
  deodorizing effect, 78

Vanilla, 2, 66, 74, 75, 92, 95, 99, 102, 104, 107, 108, 111, 112, 113, 126
  effect on the sweet sensation, 74
Vanillin, 75, 81, 175, 176
  antimicrobial effect, 175, 176
  deodorizing effect, 81
Vibrio parahemolyticus, 165
Violaxanthin, 15

Wasabi, 178, 179
Water soluble coating spice, 56, 58
Water-insoluble spice, 56, 58
Weight control effect, 149–155
White pepper (*see* pepper)

Zeaxanthin, 15, 17, 18
Zingerol, 9
Zingeron, 142
  absorption, 142
  catecholamine secretion, 142